Tsunami!

Tsunami!

Walter C. Dudley · Min Lee

A Kolowalu Book

University of Hawaii Press • Honolulu

Manufactured in the United States of America

93 92 91 90 89 88   5 4 3 2 1

Library of Congress Cataloging-in-Publication Data

Dudley, Walter C., 1945–
Tsunami!

(A Kolowalu book)
Bibliography: p.
Includes index.
1. Tsunamis—Hawaii. 2. Tsunami Warning System.
I. Lee, Min, 1943– . II. Title.
GC222.H3D84 1987 363.3'49 87-19070
ISBN 0-8248-1125-9

Book design by Roger Eggers

To the people of Hawaii
who experienced the terror
of the great waves.

CONTENTS

FOREWORD

THE OCEAN is always with us in the Hawaiian Islands. It surrounds us, brings us rain, gives us food, and provides a never-ending source of pleasure and beauty. But its power must be recognized. Those of us who live in Hawaii know the importance of the ocean to our way of life. We must also acknowledge its might.

Tsunamis are a manifestation of the ocean's power that should never be forgotten. In 1946 and 1960, people in the islands, and those in Hilo in particular, were reminded of the giant force latent in the sea. Many will remember with me the shock of the devastation, the sorrow at the losses.

Although accounts of these tsunamis were published soon after the events, this is the first time that the subject of Hawaiian tsunamis has been treated in a comprehensive way, revealing the background facts and expressing the human reality.

Min Lee has gathered together many amazing stories of the major tsunamis, and they make fascinating reading. But this book is more than a collection of tales. Walt Dudley uses his knowledge to set out clearly and accurately the why and how of tsunamis. He explains the development of the Tsunami Warning System and the way in which it works today.

The events of the evacuation during the tsunami of May 7, 1986, underlined the fact that everyone needs to be informed about tsunamis and aware of their destructive capability. The practical usefulness of this book is one of the ways it appeals to me—the other aspect is the good reading it provides.

We read about the people of Hawaii in a time of great stress. Their stories illustrate the island way—its strengths, its humor, its readiness to accept what fate has to offer and to start again. In 1946 and 1960, my company, Hawaii Planing Mill, Ltd. was levelled by the waves.

Like many other businessmen in Hilo we rebuilt, each time a little farther from the ocean!

In this book the authors show us how natural forces act and how people react; we can learn and enjoy at the same time.

Robert M. Fujimoto
Hilo, 1986

PREFACE

TSUNAMIS remain a true and constant threat to the people of Hawaii. The Tsunami Warning System has done much to alleviate this danger, and, if the warnings are heeded and understood, people need not live in fear of the waves. It is with this idea in mind that we have written this book. It is our hope that the personal experiences shared herein, together with the explanations of how these giant waves behave, will help those in Hawaii to be better prepared for the next tsunami.

In 1981, Min Lee began assembling the stories told in this book. In collecting them she met with great help and kindness from many men and women. It was a pleasure and a privilege to talk with the contributors not only about their tsunami experiences but also about many other aspects of life in Hawaii past and present. We are grateful that they agreed to share their often painful memories with us, and so with everyone.

Many thanks for the stories told to us by: Uniko (Yamani) Aoki, Charles (Chick) Auld, Tommy Crabbe, May and Lofty Cook, Robert (Bobby) Fujimoto, Bunji Fujimoto, Takeo Hamamoto, Jim Herkes, Kapua Heuer, Fusayo Ito, Bill Jensen, David Kailimai, Frank Kanzaki, Fusai (Tsutsumi) Kasashima, Evelyn Miyashiro, Claude Moore, Li and Millard Mundy, Yoshiko (Charlie) Nakaoka, Robert Napeahi, Daniel Nathaniel Jr., Herbert Nishimoto, Sushisako Okamoto, Tom Okuyama, Mark Olds, Paul Tallett, Yoshinobu Terada, Josephine Todd, Leslie Waite, Charles Willocks, Masao Uchima, and Mrs. Hideo Yoshiyama.

In addition, Richard Sillcox, geophysicist, and Gordon Burton, geophysicist-in-charge, Honolulu Observatory, Pacific Tsunami Warning Center, were very helpful in explaining and demonstrating the operation of the Tsunami Warning System.

George Curtis of the Joint Institute for Marine and Atmospheric Research gave us invaluable aid in understanding the many tsunami research and monitoring programs.

We wish to thank Helen Rogers and Thomas Hammond for their help in researching various aspects of tsunamis.

Numerous scientific research papers provided information for this book. These papers are listed in the notes at the end of the book. We express our thanks for the careful work of the dedicated scientists who studied the tsunamis described herein.

We have made every effort to be as accurate as possible in reporting the events described in this book. We know that there are many stories that remain untold.

CHAPTER ONE

With No Warning: The 1946 Tsunami

IT WAS 1:59 A.M. in Hawaii when the ocean floor off Alaska began to tremble. The movement of the sea bottom thousands of feet deep on the northern slope of the Aleutian Trench was sending shock waves through the earth—a major earthquake was occurring.

The men on the night watch at the Coast Guard lighthouse at Scotch Cap on Unimak Island 60 miles to the northwest were going about their routine duties when they began to feel the tremors. About 20 minutes later, giant waves 115 feet high crashed ashore. The lighthouse literally ceased to exist. No one would ever know what the Coast Guardsmen experienced; there were no survivors. It was April 1, 1946—April Fools Day!

Meanwhile in the Hawaiian Islands, at the laboratories of the U.S. Coast and Geodetic Survey located on the campus of the University of Hawaii in Honolulu and at the Hawaiian Volcano Observatory at Kilauea on the island of Hawaii, the sensitive earthquake-measuring devices known as seismographs had recorded the Aleutian earthquake a few minutes after it had occurred. What no one in Hawaii knew was that the waves from the Aleutians were spreading out across the Pacific; the Hawaiian Islands lay directly in their path. Although Honolulu was over 2,300 miles from the site of the earthquake, it would take less than five hours for the waves to reach Hawaii.

The first wave reached Honolulu on the island of Oahu just after 6:30 A.M. It would reach the island of Hawaii just before 7 A.M.

Laupahoehoe

It was just another Monday morning in paradise for Frank Kanzaki, a young teacher at Laupahoehoe school on the Hamakua Coast of the

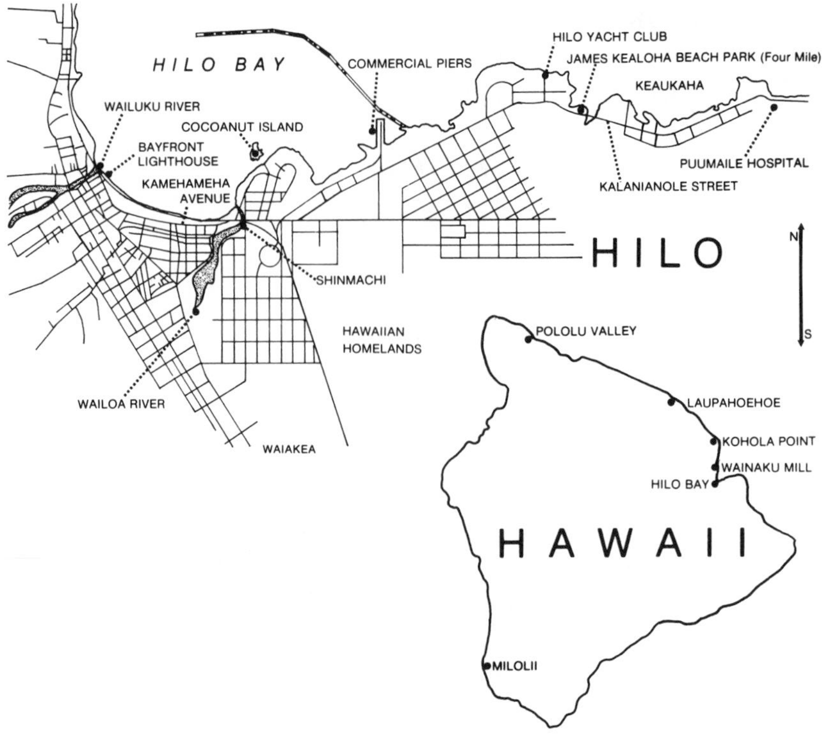

Figure 1.1 Map of the island of Hawaii and the Hilo area showing selected locations mentioned in chapter one.

island of Hawaii (Figure 1.1). As he strolled from his ocean-side cottage to the home of his friend, Peter Nakano, he looked around at the familiar scene and breathed in the early morning air. It was ten minutes before 7 A.M. The surf was high, breaking on the *laupahoehoe* lava rocks that gave the small peninsula its name. The flat expanse of grass behind the seawall was used as a baseball field, where the students from the early buses were playing and talking. Every morning Frank ate breakfast with the Nakanos. He smiled as he entered and took his place at the table between the two girls, Christine and Stella, and opposite Peter and his wife and baby. Through the window he could see the ocean—the waves a little higher now—as well as some of the students on the seawall and on the shore.

At that moment the day ceased to be like any other day.

As Frank glanced out at his carefree pupils, he saw behind them a wave that at first seemed big, then gigantic. Jumping up he seized

Christine and Stella and pushed them to the back of the house, toward the cliff. As he reached the back door, the water shattered it. The wave swept away the stilts on which the house stood and carried the cottage for a few seconds. Then the building began to disintegrate. The ceiling collapsed. Frank found himself struggling in the water and knew that he could not stay afloat while both arms were encumbered. He opened one hand—Stella was carried away. Now that he had one arm free, Frank was able to support Christine and keep his head out of the water. As the sea ebbed back to its shore, he discovered that one foot could touch firm ground. He braced his legs and prevented himself from being pulled out to sea by the receding wave. Frank was aware of Christine next to him. He raised his head and saw Peter nearby, his arm trapped between a tree and a rock. Of Peter's wife and baby there was no sign. Frank's gaze swept the area. In a space of ten minutes the scene had changed completely and forever. The wave had deposited him at the far side of the baseball field; along the edge were smashed cottages, lumber, tree branches, and the collapsed grandstand. Among the wreckage, people began to stir and call out. In the ocean were many heads, bobbing in the heaving sea.

Many of the people washed out to sea were students who had been playing near the seawall or on the shore when the wave struck. Like Frank, they had thought this just another Monday, thought that the only excitement would be an April Fool trick.

On the early school bus that carried Bunji Fujimoto to school that morning, the students had been well aware that it was April Fool's Day. Laughter of disbelief greeted the information volunteered by some, as the bus came down the hill, that the sea had receded and left the ocean floor bare. When the doubters were finally persuaded to look from the bus window, however, they discovered it was not a joke. Excited by this unusual event, but completely unaware of its significance, some left the bus as soon as it stopped and ran to the seawall. Others, a little more cautious, or less curious, stayed on the grass. Bunji was one of those who kept his distance from the ocean. As he watched his more daring classmates and his brother, he saw the water rise over the seawall. At first he did not realize the danger. It was not a crashing, breaking wave, but an ever-rising, ever-encroaching wall of water that flowed without stopping. Scared now, he turned and ran across the grass toward the higher ground. Other students fled behind him, and he noticed a member of the basketball team who made good use of his long legs, for he had been on the ocean side of the seawall when the wave began to surge; with his natural speed enhanced by terror, he achieved safety. Some were overwhelmed by the wave but not dragged out to sea because they became stuck in the bushes. Others, like Bunji's younger brother, were swallowed by the rushing sea and never seen again. Still others were blessed with a special fortune: although they had been tossed into the sea, they would live and remember the experience with awe.

~

Herbert Nishimoto would never forget what happened to him on that April Fool's Day. A tenth-grade student, he lived at Ninole but had spent the weekend "camping" with fellow student Mamoru Ishizu in the cottage that Frank Kanzaki shared with Fred Cruz. Herbert was alerted to the situation by the cries of a friend, Daniel Akiona, who ran past all the cottages shouting "Tidal wave!" Daniel's family lived in a house on the point opposite the boat ramp. He and Herbert watched the waves come and go, the second of which demolished an old canoe shed by the shore. Then Daniel urged that they should go to his house. The third wave took longer to gather than the other two had, but it looked terrifyingly huge. Desperate for refuge they dashed toward Daniel's home, but as they reached the door the whole building collapsed. Herbert was lifted by the wave, sucked into the ocean, and deposited on the reef. As he gasped for breath and stared around him he saw his friend Mamoru floating on a log and one teacher floundering in the water. Herbert tried to remove his blue jeans, but they were so tight around the ankle that he had managed to remove only one trouser leg when another wave arrived. As he dove under the wave the heavy pants, trailing from one foot, caught on the reef and he was battered to and fro. When his confusion cleared he found himself floating amidst debris, accompanied by sharks! Fortunately, the luck that had accompanied him to this point did not desert him. A section of flooring from one of the cottages floated nearby; he heaved himself onto it and lay wondering what would happen next.

Herbert's feeling of stunned shock and fearful anticipation were being shared at that moment all along the island's eastern coast. The April morning had turned from unremarkable to unbelievable, and nowhere more so than in Hilo.

Hilo: The Bayfront Area

From their home on Pukihae Street, situated atop a 30-foot-high sea cliff overlooking Hilo Bay (see Figure 1.1), Kapua Heuer and her family saw the events unreel before them. Kapua was busy preparing for the day's activities when suddenly one of her daughters asked, "Mommy, what's wrong with the water?" They all went to the cliff's edge at the end of the yard and saw that the seafloor was becoming exposed as the water withdrew. Far out at the breakwater the outward flow met an incoming wave, and the whole mass of water rushed toward the shore. Instinctively, Kapua and her daughters stepped back—just in time. As the wave collided with the sea cliff, water splashed over the tops of their coconut trees. Then the crash of the arriving wave mingled with the sound of walls and buildings being crushed as the wave struck downtown Hilo. The ocean was filled with debris and people who struggled in the waves that continued to flood into the bay. At each retreat of the water more flotsam filled the bay, accompanied by a loud sucking sound as if the ocean drank Hilo's offerings. Safe atop the sea cliff, but unable to help, Kapua watched the struggles of the victims and the destruction of the bayfront.

On board the SS *Brigham Victory,* First Mate Edwin B. Eastman was all too aware of their danger. The ship's cargo included 50 tons of dynamite and the volatile blasting caps that could set it all off. Thinking that the lines would hold them, Eastman had not been too concerned when the water first began to drain from the harbor. After the first wave, however, he changed his mind and ordered the engines to be started. By the time the biggest wave hit they were able to use their power to help keep out of trouble, but as they maneuvered they witnessed with horror a stevedore on the pier being engulfed by the monstrous wave (Figures 1.2 and 1.3). They finally managed to avoid the reefs and get past the breakwater to the open sea. But during the onslaught of the waves they were trapped in the bay and tossed about—part of the horrifying scene presented to shocked spectators.

From the vantage point of the highway bridge on the Wailuku River, Jim Herkes had his own perspective on the event. Just home from the army, he had been giving his brother Bob a ride from their home to Hilo Intermediate School. As they drove over the bridge

Figure 1.2 Stevedore (see arrow) on the commercial pier in Hilo, victim of the largest wave to strike Hilo harbor.

they noticed the exposed river bottom, and Jim knew that a tidal wave was imminent. Fascinated by the thought, he parked his car and went back to the bridge to watch. Like Kapua, Jim and Bob watched in amazement as the harbor drained. When the waves swept in, a span of the railroad bridge was carried upriver, pulled back toward the ocean, and then deposited on the small island in the Wailuku known as Maui's Canoe (Figure 1.4). Jim could hardly believe what he had seen from his safe position high above the churning water.

~

Daniel Nathaniel Jr. (known as Baby Dan to his friends) soon discovered that the river's edge was not the place to be on this April morning. Sometime earlier his cousin Alona had called to say there would be a tidal wave—she had seen the water recede from the bay. "April Fool" was his first reaction—like that of so many others. He decided, however, to see for himself, and went to the mouth of the Wailuku. He had seen small tidal waves before, yet realized with a jolt that this was to be a very different experience. As he stood on Shipman Street, a huge wave roared into the river, surged over its banks, and lifted him several feet into the air. As it carried him for-

Figure 1.3 The scene shown in Figure 1.2 as the next wave submerged the wharf.

ward he clutched at one of the roof beams on the Amfac building, and was left dangling as the water receded. Dan was shaken but safe. His slippers were in the ocean, he was in the air, but he was alive and kicking!

Fred Naylor would not be so lucky. The 84-year-old New Zealander, a well-known local character with a red nose and convivial habits, was at the Hilo railroad station when the waves arrived. In an act of heroism, he pushed a woman and her child into the station house and slammed the door just as the wave reached him. Then he was gone, lost to the sea.

All were taken unawares, there was absolutely no warning. Some were in the wrong place at the wrong time, while others escaped without knowing until later that they had been in danger.

Bill Jensen drove into town from the Wainaku mill of the Hilo Sugar Co. (a now-derelict mill just north of Hilo) and saw that the railroad bridge had been swept away. In his ignorance, an ignorance shared that day by the vast majority of Hilo's population, he did not

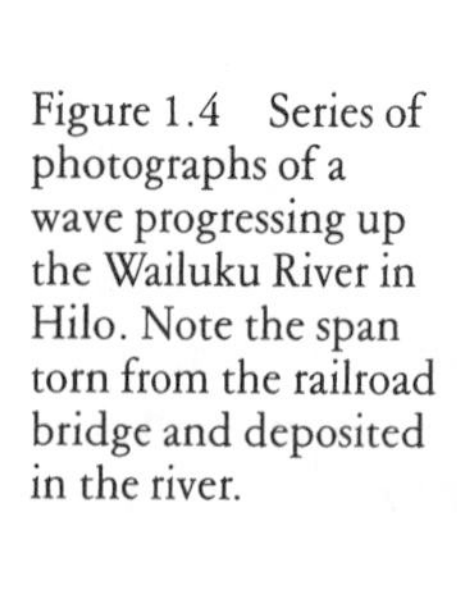

Figure 1.4 Series of photographs of a wave progressing up the Wailuku River in Hilo. Note the span torn from the railroad bridge and deposited in the river.

know that there would be more waves to come. He drove along Kamehameha Avenue, exposed to the ocean, but fortune smiled on him: his path was blocked by lumber washed from a local lumber yard so he turned *mauka* (toward the mountain). Utterly oblivious to the action of the ocean, he did not discover until later that just as he reached Keawe Street, a wave rolled over Kamehameha Avenue and would have claimed him as another victim had it not been for the pile of lumber in the street.

Bill had seen the destruction wrought by the previous wave—the businesses and shops *makai* (on the ocean side) of Kamehameha Avenue heaped into matchsticks on the *mauka* side. Up Mamo Street too, the waves brought terror. In the secondfloor apartment over the Okamoto store, Sushisako Okamoto looked out of the window to see the wave tower above the downtown bandstand. But her family was among the most fortunate—they suffered financial loss, but not personal tragedy.

Hilo: The Shinmachi Community

The greatest destruction and loss of life occurred on the coastal strip running along the bayfront and to the north of the Wailoa River. In the area that is now Wailoa State Park, there existed a close-packed, close-knit community known as Shinmachi. Founded at the beginning of this century mainly by Japanese immigrants (*shinmachi* is the Japanese word for new town), it was bustling on April 1 with the same activities that were occupying other islanders—getting ready for school and work.

In the Tsutsumi household, next to the Wailoa River, 15-year-old Fusai looked up in surprise as her sister rushed through the door, "Come see the river! It goes so fast!" Together the two girls went to look and were confronted by several boys pedalling their bicycles as fast as they could go. "Tidal wave! Tidal wave!" The sisters ran inside to warn their brother. He laughed at their simplicity, "They trick you —April Fool!" In that very instant the wave arrived with a symphony of crashes. Through the window they could see the river full of people and debris, all mingled together as the water flowed back to the ocean. They heard their neighbor, Mr. Sakido, calling to them to go to the Coca Cola bottling plant, a two-story concrete building. They ran through his house and took refuge on the second floor of the factory with many other Shinmachi residents. All of these people were to owe their lives to the fact that the manager had arrived early and the door was unlocked.

Hiromo Tsutsumi (Fusai's brother) had decided to run back to their house to see what he could save. The adults prevented the younger people from leaving the Coca Cola building, but Hiromo was old enough to make his own decisions. Once back at his house, however, he could not decide in the excitement of the moment what to take away with him. Looking around, he caught sight of a *furushike* (a Japanese cloth used for carrying small articles) full of family photographs and pictures. He seized it and ran back to the safety of the Coca Cola building. The next day his father protested that he had not saved the considerable amount of money in the house. But in the years to come the mementoes were of far greater value than mere money. Many survivors echoed the Tsutsumis' feelings: the sentimental loss of personal effects was of more importance in the long term than the destruction of buildings or the loss of money. At the time, however, those in the Coca Cola building were thinking only of survival.

Masao Uchima was there too. He had been in bed when he heard his father shout "Tsunami!" Water was everywhere, it seemed to him that the island was sinking. His father shouted for everyone to go to the Coca Cola building—they reached it before the next wave arrived. Thus Masao became an observer, not a victim. From the roof of the building he saw a 15-foot wave roll in, followed by an enormous wall of water that stretched right across the bay. This wave obliterated most of Shinmachi, except the building on which he stood.

~

It was after the second wave that Yoshinobu Terada found himself in the river. When the water began to recede the first time, his father had sent him to look at their boat. As he was returning home he saw a 4-foot wall of water rolling toward him up an alleyway. He leapt onto the nearest porch: from all around came the crash of structures crumbling—a row of houses telescoped and the buildings rammed together. The house on which he was standing was carried into the river and began to sink. He jumped into the water, where there were many other people and a vast amount of debris; he heard the cries of people struck by swirling beams, appliances, doors, and roofs. All was turmoil and time stood still. Fortunately, however, Yoshinobu saw his brother's surfboard floating by him in the shambles. He climbed onto it, happy to be raised above the rubble with a means of reaching shore. He paddled to land just before the next wave arrived and went to look for his family in ravaged Shinmachi.

The Shinmachi township extended to the Hilo Ironworks and the Wailoa River. On the other side of the river was Waiakea town. Hilo had begun its commercial expansion in this area with the building of the railway terminus in 1899.

Waiakea

In 1946 there was a gas station and store at the crossroad next to the bridge, as well as a newly opened cafe. The cafe owners, Evelyn and Richard Miyashiro, had started their business in January. Now they were on the bridge with the other Waiakea residents, waiting to see the big waves. There had been other tidal waves before, when the water had risen only a few feet. With the arrival of the first two waves excitement rose among the watchers. The third wave, however, was immense—there was no time to run. The Miyashiros were there when the massive wave swept over the bridge, and, amazingly, they were still there after it had receded. Evelyn Miyashiro, eight months pregnant with her first child, clung to the concrete uprights of the bridge, her husband beside her. Coconut trees were flattened, the building housing the Kitagawa auto dealership collapsed, and the Miyashiro's cafe was flooded to a depth of 3 feet—but Evelyn and Richard were still alive!

~

Unlike the Miyashiros, the Hideo Yoshiyama family suffered the loss of several loved ones. Unaware of their peril, Mr. and Mrs. Yoshiyama and their young son Alan went outside their home on the bank of the Wailoa River to watch the waves arrive. After the first waves, several people shouted that a "big one" was coming. Mrs. Yoshiyama returned to the house to warn her parents, Mr. and Mrs. Fukui. Meanwhile, the next wave arrived with great force. Hideo was swept by the wave into the river—his son Alan torn from his grasp.

Inside the house, Mrs. Yoshiyama was having a difficult time convincing her parents to leave the building—her father was reluctant to leave his plants. Mrs. Yoshiyama gave up the attempt to persuade him and ran upstairs to the living room with her mother and her niece, Alison. The wave invaded the house and carried them across the room in a jumble of lumber and furniture, Mrs. Fukui clinging to her daughter with one hand and clasping her grandchild desperately in the other. The powerful water pulled at her as it moved back toward the ocean. Her daughter was stuck fast in the debris, but the inexorable pull of the flowing water must have made Mrs. Fukui fear that they would all be swallowed in the muddy torrent. She snatched her hand free. Then she and Alison were wrenched down the bank and disappeared into the hungry sea.

Mrs. Yoshiyama still feels sorrow at the loss of her three-year-old son and her mother, but feels fortunate that her two daughters, who had been on their way to school, escaped without injury. Her father was swept from the basement room where he had remained with his plants, but was washed up alive on the bank of the river.

Some residents of the Waiakea district were more fortunate than others. When Josephine Todd looked out into the bay that morning and saw the water recede, she knew what was going to happen. She sent her children with her niece to her sister's home on Kaumana Drive. Meanwhile, she busied herself moving possessions from her home to the house of her niece, which was nearby but on higher ground. Her idea was to save them from a soaking. Like others in the Waiakea district, she had seen tidal waves before, but only small ones. Familiar with the unpredictability of the ocean, she kept a watchful eye on it as she made the trips between the houses during the periods the water was receding. Between trips, Mrs. Todd counted seven waves. When the eighth arrived it was a towering black mass, higher than the trees on Cocoanut Island. She ran to the car to make her escape. As she drove away she could hear the wave smashing her house.

Millard Mundy was also one of those who tempted fate, but escaped unharmed. In 1946 he, his wife Li, and their two children lived in a house on the shore opposite Cocoanut Island. Employed as a teacher, he was in the bathroom shaving to prepare for school when he saw his neighbor, Mrs. Richardson, letting all her chickens out. Then another neighbor, Rose Chock, shouted that a tidal wave was coming. Mr. Mundy saw that the water in the bay had drained out, leaving the bottom bare. He hurried his pregnant wife and two children into the car in the driveway, but decided that he could not miss the opportunity to take photographs of such a fascinating natural

Figure 1.5 Giant wave washing over Cocoanut Island in Hilo Bay.

phenomenon. He stood by the water's edge as successive waves came, snapping his photographs (Figure 1.5). The first few waves did not discourage Mr. Mundy—although Cocoanut Island was inundated, the nearby shore where he stood did not flood. The next wave, however, was a terrifying sight; as it crashed onto the breakwater, boulders the size of automobiles were flung into the air. At that moment, Millard Mundy decided that he was in the wrong place. As he accelerated out of the driveway, the house next door was carried out into the sea—colliding with his home as it swept past.

The devastation caused by the waves continued along the coast, wreaking havoc in Keaukaha.

Keaukaha

At the Hilo Yacht Club, Yoshiko "Charlie" Nakaoka and other staff members were in their quarters behind the club building when the water rose up to the level of the steps. It was fortunate for them all that the manager, Mr. Kennedy, recognizing the signs of a tidal wave, ushered them into his car and left at once. As they made their escape by the back roads, the great wave washed away all buildings on the site of the Yacht Club. At the same time, it crashed through a private home next to the Yacht Club. Unfortunately, the owner, Betty Armitage, had not been as quick to escape as the Yacht Club staff. The waves engulfed her car and, when she tried to leave the vehicle, she was swept out to sea by the receding waters.

Farther into Keaukaha, Leslie Waite was enjoying an early-morning breakfast with his wife and two children. Dr. Waite had just returned from Honolulu, bringing with him some special bakery rolls. The family was sitting at the breakfast table when the maid hurried out of the kitchen. She didn't have the rolls. Instead she announced that the yardman, Nishiona, was at the door. "He say the ocean look funny." As she spoke, she looked out the window, and then screamed, "Tidal wave!" The Waites failed to be alarmed. The maid was only 16 years old and tended to be very excitable. At that point the yardman burst into the room with the news that the narrow strip of road connecting Keaukaha to the main portion of Hilo had been flooded by the ocean. Together, they rushed to pile into the car—the four Waites, the maid, Nishiona, and the one dog they could find. But as he prepared to drive off, Dr. Waite had second thoughts. If the road had flooded already, it might not be safe. Instead, he decided they should walk to the high ground occupied by the navigation towers, a series of poles used as navigational aids by aircraft destined for the Naval Air Stations or Hilo Airport. When they reached the towers, Mrs. Waite left her baby with the maid and ran back toward the house with the intention of salvaging her spectacles and her purse. As she drew near to her home she heard the roar of the approaching wave, and decided to retreat to safer ground. Meanwhile, her husband had gone to help some elderly neighbors and was across the street from his home when the roaring waters struck it. He watched helplessly as the next wave scattered the wreckage in the street.

Other residents from the area were gathering by the navigation towers. The Willocks family had been getting out of bed when the first waves arrived. Charles Willocks, manager of the Hilo Iron Works, opened his back door to see four feet of water at his back steps and big fish flip-flopping in the grass. He immediately collected his wife, sons, and maid, and ran to the cars—only to find that they had been jammed together by the wave. So they ran on, up the steep driveway of the house opposite, across the back yard, and inland as fast as they could. They did not stop to watch the waves destroy their house and deposit the debris all over the driveway they had used for their escape.

The driveway belonged to Lofty Cook. The deep hollow in his front yard became the depository for much of the lumber carried by the wave. The Cook family made their escape in the same direction as the Willocks. From their house (which was built on a rocky knoll and, as a result, survived all the waves) they had watched the first waves roll in before deciding to evacuate.

~

Farther along Kalanianaole Street, Paul Tallett had been out since before 7 A.M. waiting for the newly inaugurated bus service. Dressed for work in his suit and tie, he waited by his front door. Looking out over the ocean, he noticed that the water was an unusual color, a dark green. He went toward the shore for a better look, and as he did the water receded—tumbling huge boulders in the backwash. He hurried back toward his house, passing the home of Ernie Fernandez, who was out raking his yard. "A big sea is coming," he told him. Ernie and his family joined the Talletts at their home, which was built on a raised outcrop of rock. From there they saw another wave advance as a mountainous swell which did not break, but surged onto the shore. Other neighbors joined them on this high spot of land. Paul thought of the Kais and knew they would need help, as Mrs. Kai was blind. With a rope tied around his waist, Paul tried to reach the Kai home but the flood was too deep. The sea engulfed the land with a tremendous roar. As the oceanside fish ponds that had been emptied by the withdrawal of the sea were refilled by the incoming wave, a tremendous boom filled the air. The watchers saw the Kwock home demolished. The Kai home was carried out to sea, levelling coconut trees as it went; with the next wave it was lifted and dropped onto the rocks where it burst apart. The people gathered in the Tallett yard were aghast. Would the whole island drown? Before the arrival of the next wave, Paul urged his assembled neighbors (more than 20 people) to evacuate the area. Along the route, Paul discovered the young Mrs. Nuhi and her baby stranded in a tree. Working quickly he helped them down, and led them to higher ground. Another wave came as they were pushing their way through the trees, but they were far enough from the shore to be safe.

~

All the ponds in Keaukaha emptied and filled many times that morning. At the Richardson estate, the water passed straight through the house but left it intact. Across the street, Charlie "Chick" Auld had noticed the road was flooded when he got up that morning. He and his son climbed to the roof of their house where they watched the progress of the giant waves. One wave rushed in, accompanied by crashes and roars. The next ran up their driveway and took the car from their garage. The wave after that carried their car to Puhele Street, but the house remained undamaged—the Aulds secure on its roof. Mingled with the noise of the waves, they heard the cries of those who had been in the wave's path.

Such cries were echoing up and down the coast. The big waves had all arrived by now. It was nearly 9:00 A.M. Monday morning, the air still balmy. But the coastline was devastated, laid to waste. For many victims of the giant waves, life was changed forever; for others, life was over.

Rescue Efforts

Now the massive task of rescue was about to begin. Some who had been involved in the full fury of the waves would survive, others would not. But heroic efforts were made on their behalf, by individuals and by organizations.

The Hawaii County Fire Department mobilized very soon after the waves. Robert Napeahi had been on duty at the Central Fire Station, located then, as it is now, at the corner of Kinoole and Ponohawai streets. Just after 7 A.M. people came running up Ponohawai Street, shouting that the Okano Hotel had been flooded by the waves. The firemen went to the roof of the fire station and from there saw the effect of the biggest wave. They could not see the wave itself, but as they watched, the Okano Hotel collapsed.

The firemen were quickly divided into teams. Bob Napeahi took his section to Waiakea. As they made their way from the fire station, they saw destruction everywhere. Buildings on the makai side of Kamehameha Avenue had been washed across the street. The Cow Palace and the Hilo Theater alone had survived on the ocean side. By the Hilo Iron Works the firemen saw a man stranded at the top of a coconut palm, afraid to attempt the descent. When threatened by an oncoming wave, he was able to scale the tree, but now was stuck at the top. People had taken refuge on rooftops, on planks floating in the river, and in skiffs. Many had been caught unaware in their homes, and were trapped when the buildings collapsed.

Bob helped to rescue one couple through the roof of their home, using his weight to break through the ceiling so that they could climb out. Those rescued were taken by truck to Hoolulu racetrack or to the railroad freight depot, where they were tended by Red Cross workers. But beside the living were many dead: a young woman drowned in her car, a mother and her baby crushed between a large banyan tree and the broken slab of a wall, others floating dead in the river. These bodies were taken by fire truck to a central location to await identification. It was a long and mournful day, with all possible agencies and able individuals involved in the rescue effort.

Those who witnessed the event knew that there was much to be done. Daniel Nathaniel, left swinging from a rafter when the wave receded, dropped safely to the ground. He discovered with joy that his brother had survived also, even though the wave had knocked him down. Five men hastened along the street, looking for their wives. The women had run into the Hilo Meat Market—which had then been flooded—but they were able to break out of the back door ahead of the on-rushing water. At the nearby Amfac building the Nathaniel brothers helped the manager to stack up the goods that had been knocked down by the waves. Fortunately, because the waves hit early, there had not been very many people in the shopping area.

~

Kapua Heuer, who had watched the terrifying chain of events from her cliff-top home, waited until she was sure that the waves had finished coming in, then decided that she would be needed at her workplace, the Ordnance Depot at the Naval Air Station. As she went into the street she was confronted by the amazing sight of one of her neighbors, 90-year-old Mr. Spark, stark-naked in the middle of the railroad track. He had been washed out of his bathtub and now sat considerably astonished—but unharmed.

When Kapua arrived at work, her boss asked her to go with a coworker, Phil Brown, to see what had happened in Keaukaha. There was particular concern about the fate of patients in the tuberculosis hospital at Puumaile.

They drove along Kalanianaole as far as Awili Store, where the road was blocked. A young boy came riding along on a horse, so they commandeered it. They rode to the home of Kapua's friends, the Codys. Mrs. Cody was in the hospital for the delivery of her second child. At the house Kapua found Sam Cody and asked, "Where's Meredith?" (Sam's three-year-old daughter). He stood stunned, then said, "She's gone." During the big waves he had opened the door and told the maid to run to safety with the little girl. Neither the maid nor the girl were ever seen again.

There was nothing to be done at the Cody house, so Kapua and Phil continued with their investigations. At "Four Mile" (James Kealoha Beach Park) the road had been washed out, so they backtracked around Desha Street, left the horse, and swam across Lokoaku Pond. It was not an easy swim—the water was rough and full of debris. They rested for a while by the ruins of the Carlsmiths' house, taking refreshment from a papaya and a bottle of wine they found floating in the water. Then they continued toward Puumaile hospital and met some of the patients, many of whom had been confined to bed for years, straggling along the highway. Some 300 tuberculosis patients had been marooned in the hospital when the waves had washed out the road and inundated the grounds, and it was feared that subsequent waves might destroy the buildings. With an idea now of what needed to be done, Kapua returned to the Naval Air Station. Lt. Commander Barber organized the evacuation of those in Keaukaha, and mobilized the Marines to employ tanks to bridge the water at Four Mile and use rubber dinghies to ferry people across. Soldiers, sailors, and civilians all lent a hand to carry the patients to safety at the navy barracks where doctors, nurses, and corpsmen provided necessary aid.

Among those ferried across the waters at Four Mile were all the people who had taken refuge at the navigation towers. Since they had gathered there, they had been trying to keep up their spirits; most remained calm, with the exception of one elderly woman who was hysterical. Dr. Waite had spent his time reassuring the older people, while Mrs. Waite, a teacher at Hilo High School, used the books that the Cook girls had with them to occupy the children.

During this period, Lofty Cook returned to his home to fetch some food, as well as some milk for the babies. When the family had left early that morning, the table was set for breakfast, including several portions of cut grapefruit. On his return he found that the grapefruit had been eaten! He never knew whether it was eaten by a mongoose, a menehune, or a hungry neighbor!

Paul Tallett and Willie Ishibashi also returned to their homes, primarily to look for some of their neighbors who had never arrived at the high ground. They were particularly concerned about the Pua family. As they approached Paul's home through the floodwater, they found the bodies of old Mrs. Nuhi, and her grandson and two granddaughters. Then they heard crying. It was Mrs. Pua; they found her battered and bruised, part of her pajamas torn off. Paul gave her his raincoat and walked with her to the high ground. Once more he returned to his house and collected as much food as he could carry, and, for sentiment's sake, took an old Hawaiian crook (walking stick) that had belonged to his grandmother.

While Paul was aiding survivors in Keaukaha, attempts at rescue were being made in all other stricken areas. In Laupahoehoe however, efforts were sadly hampered by the lack of boats.

~

David Kailimai, the superintendent of the Hamakua Mill Co., had been at his home in Kukuiau when the waves struck. About an hour later, one of his workers called him to say that there had been a "big flood" in Laupahoehoe and trucks were needed to salvage possessions. As he drove down the steep road to Laupahoehoe, he saw the floating debris and levelled buildings and knew there had been a tidal wave.

He knew also that a boat was essential to save the people—about ten of whom he could see floating in the ocean. No help could be expected from boatowners in Hilo, since it seemed all too likely to David that their need for boats would be equally great, and that many vessels would have been damaged or destroyed. His own boat was on the other side of the island in Kona, but he did have the outboard motor at his house. The only boat in the Laupahoehoe area was a sailboat belonging to a Mr. Walsh. At first Mr. Walsh was not willing for his boat to be used: the sea was still rough and full of debris, the wind was strong, and it seemed likely that damage would occur. Moreover, the boat would have to be cut to accommodate David's motor. It took much persuasion—by the time the boat was launched it was after 2 P.M. The action of the wind and waves had scattered the people David had seen earlier. He set off to search for survivors, accompanied by Libert Fernandez, a young boy named Masaru, and a young soldier, Francis Moku Malani who was home on compassionate leave. Dr. Fernandez was anxious for the safety of his fiancée, a young teacher. First they found two boys, students from the school; then a seaplane, flying overhead, directed them to where it had dropped a rubber raft. On the raft they found Marsue McGinnis, the fiancée of Dr. Fernandez. Of the others who had been floating in the water there was no trace.

Marsue McGinnis lived with three other teachers in the first of a group of faculty cottages. She was the only one of the housemates to survive. In the second cottage, the four student teachers were more fortunate: all of them lived, although one woman, Evelyn (later Mrs. Tommy Crabbe), had the tendon in her leg severed by the broken glass of a bookcase.

In the next cottage four more young women survived, including Uniko Yamani. Following a warning shouted from a neighbor, Uniko looked out of the window and saw "an ugly black wave." In her terror, she found herself climbing the room divider, screaming out and calling for her mother. Two of her housemates cowered on the sofa, and the third rushed from the bedroom in response to Uniko's screams—just in time, for the bedroom walls splintered as the wave struck the cottage. In the afternoon she went back to the ruins of the cottage and, although the house had been torn in half, the living room remained almost intact. Still in a state of shock, she entered the house because all day she had been worrying about her panty girdle, hanging in the bathroom exposed to view. Uniko found her girdle and took that and her diploma from the ruins. She doesn't recall why she chose those things over the rest of her possessions.

The cottages were found wedged on the wall of a piggery—an action that probably saved them from complete destruction. It didn't save the pig, however; he had disappeared into the sea.

Herbert Nishimoto remained stranded in the ocean on his improvised raft. After a period of time he sighted two fellow students, Takemoto and Kuhuki, and pulled them aboard. At about 1 P.M. the seaplane that had dropped the raft to Marsue McGinnis dropped one to them. They had no fresh water to drink, but did recover some coconuts they found floating in the water. They decided to paddle away from land to avoid the large masses of debris. Darkness fell. Between spells of dozing, they tried to locate the coast, but only saw one light. By morning they had drifted more than 10 miles south to Kohola Point. Another plane flew over them, attracting the attention of a young girl on shore who was on the way to her grandmother's house. She saw the raft and ran to tell a group of cane workers who were coming home for lunch. Two strong swimmers went out to tow the raft, whose occupants were too weak to swim to shore themselves. It was a happy ending for Herbert and his companions.

But for many others on the east coast of the island of Hawaii, that day and many days following were spent counting the cost of the sudden disaster.

~

On the five main Hawaiian Islands, 159 people were killed by the waves. But the island of Hawaii was by far the hardest hit. In Hilo alone, 96 people lost their lives to the sea. At Laupahoehoe, 16 school children and 5 of their teachers were lost. Of the 159 people killed only 115 bodies were ever found. The enormity of the loss was brought home to the people of Hilo by the sight of bodies lined up in the streets outside Dodo Mortuary. Kapua Heuer was one of the many who went there on the grim mission of searching for a missing friend or relative. She remembers still the horror on the faces of the dead.

There were so many bodies that some had to be moved to an icehouse for storage. They were stacked in such a way that when it was time for burial, the workers found that the corpses had frozen together—necessitating the gruesome task of chipping them apart with ice picks. Such episodes heightened the atmosphere of unreality that the disaster had cast over Hawaii.

Yet the toll in human lives could have been much worse—the timing of the tsunami was one bit of fortune. By 7 A.M. most people, although still in their homes, were at least awake. Had the first wave arrived an hour earlier, many people would have been caught asleep in their beds. If the waves had arrived an hour later, when people had left home for the day, downtown Hilo would have been filled with early-morning workers and shoppers. In either case the loss of life would have been much greater.

The timing was irrelevant, however, in terms of property damage. Almost 500 homes or businesses were totally destroyed and another thousand severely damaged; the cost of the destruction totaled an estimated $26 million. Structural damage included buildings, roads, railroads, bridges, piers, breakwaters, fishpond walls, and ships.

Frame buildings situated near sea level suffered the most damage: some were floated off their foundations nearly intact, while others were totally demolished where they stood. Many of the two-story frame buildings suddenly became one-story structures when their ground floors were washed away and their top stories were left to rest on the foundation.

The densely built bayfront business district in Hilo was almost totally demolished (Figure 1.6). Nearly every house on the side of the

Figure 1.6 Damage to the downtown Hilo business district caused by the 1946 tsunami.

main street facing Hilo Bay was smashed against the buildings on the other side, but those that absorbed the brunt of the waves saved many of Hilo's downtown stores located farther inland.

The railroads in Hilo as well as those along the northern coast of Oahu were wrecked. Railroad cars were overturned on Oahu, Maui, and Hawaii (Figure 1.7). The Hilo train station disappeared. The rails of the Hawaii Consolidated Railway were torn off the railbed and, in some cases, were wrapped around trees. In other cases the tracks were moved en masse (Figure 1.8), probably floated off the roadbed by the buoyancy of the wooden railroad ties.

Jim and Bob Herkes saw an entire span of the steel railroad bridge across the Wailuku River sheared from its supports and washed 750 feet upriver—passing under but not damaging the highway bridge on which they were standing. At Kolekole Stream, 11 miles farther north, an entire leg of a high steel railroad trestle was twisted off its base and carried 500 feet upstream. The Hawaii Consolidated Railway, already in financial straights, was forced to close down as a result of the extensive damage to its property.

At Hakalau Gulch, 15 miles north of Hilo, the sugar mill suffered

Figure 1.7 Railroad caboose washed under a building during the tsunami of 1946.

Figure 1.8 Iron rails shifted by the force of the waves of the 1946 tsunami.

Figure 1.9 The eighth wave withdrawing from Hakalau Valley. The sugar mill of the Hakalau Sugar Company was virtually destroyed by the water, which rose to a height of 20 feet.

Figure 1.10 Waves washing up Waianuenue Avenue in Hilo.

Figure 1.11 Aerial photograph of the commercial piers at Hilo after April 1, 1946. Note the SS *Brigham Victory* returned to harbor.

severe damage (Figure 1.9). Its location at the mouth of the gulch was only about 10 feet above sea level. Coastal highways also were partly destroyed, largely by undercutting. Moreover, as water flooded into the streets of Hilo (Figure 1.10) dozens of automobiles were tossed about and wrecked.

The mile-long breakwater, which protects Hilo harbor from normal ocean waves, helped reduce the impact of the giant waves, but nearly 60 percent of the structure was destroyed. Giant blocks of stone, some weighing more than 8 tons, were strewn on the bayfront beach like grains of sand. Pieces of coral 5-feet wide were wrenched from the reefs and tossed on the shore to elevations more than 15 feet above sea level.

The commercial piers in Hilo harbor were severely damaged by the force of the waves washing over them and by the debris which acted as waterborne battering rams (Figure 1.11). Many small boats, including pleasure craft and sampans of the local fishing fleet, were washed ashore and damaged (Figure 1.12).

The Naniloa Hotel, situated on the peninsula projecting into Hilo Bay, was largely left intact, although it did lose its dining room, boat house, and swimming pier. The hotel was practically the only building along the shoreline not totally destroyed or seriously damaged.

The giant waves from the Aleutians had, in places, swept more than a half mile inland. The withdrawal of the waves between crests exposed the seafloor for a distance of up to 500 feet below the normal strand line. The water attained heights ranging from an enormous 55 feet in Pololu Valley (see Figure 1.1) on the island of Hawaii to as little as 2 feet at the village of Milolii on the opposite side of the island. Photographs taken in Hilo show that the tops of breakers were more than 25 feet above the normal water level in the bay as they swept over Cocoanut Island (see Figure 1.5).

The people of Hilo had already lived through four years of world war fearing possible Japanese attack. The year 1946 should have been free of fear that Hilo might be destroyed. Yet the forces of nature—the catastrophic waves—had brought Hilo to its knees. The entire community valiantly went about the task of cleaning up and rebuilding after the disaster (Figure 1.13).

Figure 1.12 *(top)* Fishing sampan washed ashore and overturned by the force of the tsunami waves in 1946.

Figure 1.13 *(left)* Residents of Hilo, cheerful and brave, clean up after the disastrous tsunami of 1946.

CHAPTER TWO

What is a Tsunami?

As HILO dug itself out of the rubble caused by the great waves, state and local authorities began the task of trying to determine exactly what had occurred. Newspaper accounts all told of the "tidal wave," but we now know that the terrible waves that came from the Aleutians and wreaked havoc in Hawaii had nothing to do with the tides. Such enormous, destructive waves have been called "seismic sea waves" by scientists, but are now generally referred to by the term *tsunami.* Tsunami (pronounced tsoo-nah-mee) is a Japanese word meaning "great wave in harbor." The name is appropriate as these giant waves have frequently brought death and destruction to Japanese harbors and coastal villages. For more than two thousand years the Japanese have recorded the dangers posed by tsunamis; their awesome power is dramatically depicted in the famous nineteenth-century print by Hokusai (Figure 2.1).

Tsunamis have, no doubt, visited the shores of Hawaii since the islands first formed. It is almost certainly the occurrence of tsunamis that gave rise to the legends found in Hawaiian folklore of the sea engulfing the land (Malo 1951). One such story tells of a love affair between a woman who lived in the sea outside Waiakea, Hilo (an appropriate place for a tsunami legend) and the reigning king of the area, named Konikonia. The woman was lured ashore to sleep with the king, although she warned him that her family would come looking for her. It seems that her brothers were *paoo* fish, and in order for them to be able to search for her the sea would rise. Accordingly, after ten days had passed, "the ocean rose and overwhelmed the land from one end to the other" until it reached the door of Konikonia's house. Many were drowned, but "when the waters had retreated, Konikonia and his people returned to their land."

With the arrival of western missionaries in Hawaii, destructive tsunamis were chronicled in detail, often with religious overtones.

Figure 2.1 *Great Wave off the Coast of Kanagawa* by Hokusai.

Richard Armstrong, a missionary on the island of Maui, documented a tsunami occurrence at Kahului on November 7, 1837 (Jaggar 1946). It was around 7 P.M. on a calm evening, when without warning the water began to withdraw from the beach. As the beach widened to more than 120 feet and the reef became exposed, delighted natives rushed out to pick up stranded fish. A few individuals, probably having experienced such a phenomenon, concluded that the sea would soon rush in again, and fled toward higher ground. But most were caught unaware as a terrifying wall of water surged back in. One man saw the water coming into his house and, grabbing his child, ran to safety. As he turned his head to look back upon his home, he was astonished to see "the whole village, inhabitants and all, moving toward him, some riding on the tops of their houses, some swimming, all screaming with fright." The tsunami had carried the entire village of 26 grass houses—complete with their inhabitants, canoes, and livestock—some 800 feet inland, dumping it all into a small lake.

Fortunately, many of the villagers were good swimmers and

managed to stay afloat. Some even swam to safety with children, the sick, and the aged on their backs. As Reverend Armstrong summed up: "By the blessing of God, all escaped but two at this place. One of these was a mother who was carried out of the flood by her son, safely, as he supposed, rejoicing that he could aid her in such peril. But how was he disappointed when he laid her on dry ground, to find that she had been overpowered by the shock and was dead!"

At Hilo the scene was more terrible than at Wailuku; about 10,000 people had been assembled at the bay for religious instruction. After spending a long day in church services, the people had either gone home to rest, or were gathered along the shore at sunset when the sea began retreating. An English whaling vessel, the *Admiral Cockburn,* was anchored in the bay at the time and the shipmaster, Captain James Lawrence, stated that a "great part of the bay was left dry." Natives rushed down in crowds to witness the strange sight, when suddenly a gigantic wave formed and surged toward them.

The Reverend Titus Coan, who witnessed his flock's distress, stated (Bingham 1847):

> God has recently visited this people in judgement as well as mercy. . . . The sea, by an unseen hand, had all on a sudden risen in a gigantic wave, and this wave, rushing in with the rapidity of a racehorse, had fallen upon the shore, sweeping everything into indiscriminate ruin.
>
> So sudden and unexpected was the catastrophe, that the people along the shore were literally "eating and drinking," and they "knew not, until the flood came and swept them away."
>
> Some were carried out to sea by the receding current. Some sank to rise no more till the noise of the judgement wakes them.

The waves surged into the village at Waiakea, rising to 20 feet above high water. According to Reverend Coan, the sea crashed upon the shore "as if a heavy mountain had fallen on the beach."

Men, women, and children struggled in the flood, amid their wrecked homes. So violent was the suction as the sea withdrew, that even strong swimmers could make little way, and some sank exhausted. But fortunately Captain Lawrence ordered his sailors to "search for those floating upon the current," and thus some were rescued by the boats of the *Admiral Cockburn.*

The scene on shore was horrible. About a hundred houses filled with their occupants and guests had been totally demolished and washed away. "Half frantic parents were searching for their children; children weeping for their parents. Husbands running to and fro inquiring for their wives; wives wailing for their departed husbands."

At Hilo, four men, two women, and five children had lost their lives, and at a nearby village two women and a child were drowned. But as Reverend Coan stated: "Had this providence occurred at midnight, when all were asleep, hundreds of lives would undoubtedly have been lost. But in the midst of wrath God remembered mercy."

The lesson taught that day by Reverend Coan had been "Be ye also ready." And as the Reverend later said, "This event, falling as it did like a bolt of thunder from a clear sky, greatly impressed the people. It was as the voice of God speaking to them out of heaven." It was a lesson the missionaries would not let the villagers forget.

Even the salty English sea captain, James Lawrence, was impressed by the day's events. According to Coan, "he was a large and powerful man, bronzed by wind and wave and scorching sun, who had thought little of God or the salvation of his soul." But after experiencing the tsunami, "he knelt at the altar and professed to give himself to the Lord." On returning to his ship, he immediately told his officers and crew that he would drink no more, swear no more, and chase whales no more on the Sabbath!

Small waves would continue to surge in and out for a day or more. The tsunami of 1837 would be remembered both as an act of God and as a terrible natural disaster.

In 1877 such a disaster would again be visited on the people of Hawaii when, on May 9, a great earthquake occurred off the coast of South America between Peru and Chile. The first waves reached Hilo in the dark before dawn on the morning of May 10. Sheriff Luther Severance of Hilo recalled the events of the day (Hitchcock 1911):

> We have had a great disaster at Hilo. On Thursday morning the 10th at about 4 A.M., the sea was seen to rise and fall in an unusual manner, then at 5 A.M. washed up into nearly all the stores in the front of the town, carrying off a great deal of lumber and all the stone wall makai of the wharf.
>
> But at Waiakea the damage was frightful; every house within a hundred yards of the water was swept away. The steamboat wharf and the storehouse, Spencer's storehouse, the bridge across the stream, and all the dwelling houses were swept away in an instant and now lie a mass of ruins far inland.

The American whaling vessel *Pacific* was anchored in Hilo Bay in 24 feet of water at the time of the tsunami. When the sea receded the ship was left high and dry—but then the waters surged back in, spinning it round and round. All observers expected to see the vessel drag ashore, but somehow she survived. Her master, Captain Smithers—like Captain Lawrence 44 years before—sent his longboats off across the bay to save the helpless people swimming for their lives in the swirling water.

A small hospital situated on Cocoanut Island disappeared as the waves washed completely over the island. A small church at Waiakea was floated off its foundation and washed about 200 feet inland; reportedly it ". . . travelled with much dignity, tolling its bell as it went, and was scarcely injured at all, while the principal houses beside it fell in total ruin."

Another account of the events of the morning was reported in a letter from the wife of Reverend Coan (Coan 1882). As she recalled:

> I was just rousing from quiet slumber this morning, not long after five, when heavy knocking at our door hastened me to it. . . . A [giant] wave had swept in upon the shore; houses were going down and people were hurrying mauka with what of earthly goods they could carry.
>
> Houses were lifted off their under-pinning and some had tumbled in sad confusion and lay prone in the little ponds that remained of the sea in various depressed places. Riders at breakneck speed from Waiakea brought word of still more complete ruin there; the bridge they said was gone!
>
> People were wading in water where their homes had stood half an hour before, gathering up goods soaked by brine.
>
> At Kanae's place, the word was that old Kaipo was missing. Asleep, with Kanae's babe pillowed near her when the wave came upon them, she had wakened, and hastening out of the house found herself in deep water. Holding the little one above her head, she had the courage and strength to keep it safe till the mother swam for it, and then, no one knows how, the old woman was swept out to sea. . . .

Mrs. Coan also related the story told to her and her husband of the remarkable escape of a 77-year-old English resident of Hilo. "I got caught, sir. . . . I should have escaped if I hadn't gone back after my money; when I came down-stairs the roller had hit the house, and before I could get out of the door, the house had fallen upon me. I was dreadfully bruised, and you see sir, as the wave took the house inland, it kept surging about with me in it, and getting new knocks all the while." "And what of the money—was it saved?" "Oh, no, sir, it all went, six hundred dollars. It was all I had, and I am stripped now and I'm past working. . . ."

Total damage to Hilo caused by the 1877 tsunami was 5 dead, 17 badly injured, 37 houses destroyed, 163 left homeless and destitute, and 17 horses and mules drowned.

As word of the disaster spread to the neighbor islands, help would come from every quarter. Her Royal Highness Lydia Dominis, King Kalakaua's sister, came with donations from Honolulu. Long after the event, May 10 was commemorated as a day of thanksgiving for the aid received.

For many years after the catastrophe, a zinc plate nailed to a coconut tree a short distance from the beach served as a reminder of the big waves. The plate, 5 feet above ground level, marked the height to which the waves rose during the 1877 tsunami—more than 16 feet above the sea.

The 1877 tsunami would remain the most disastrous on record for Hilo until April 1, 1946.

The Mechanics of Tsunamis

Just what is a tsunami and what causes it? In the very simplest terms, a tsunami is a series of waves most commonly caused by violent movement of the seafloor. In some ways a tsunami resembles the ripples that radiate outward from the spot where a stone has been thrown into the water, but a tsunami occurs on an enormous scale.

The movement of the seafloor that causes the tsunami can be produced by three different types of violent geologic activity. By far the most important of these is submarine faulting, when a block of the ocean floor is thrust upward or suddenly drops (Figure 2.2). Such fault movements are accompanied by earthquakes. The earthquakes associated with tsunamis are sometimes referred to as "tsunamigenic earthquakes" and it is often stated that earthquakes are the cause of tsunamis, but this is not true. The earthquake does not cause the tsunami, but rather both result from the same fault movement.

Probably the second most common cause of tsunamis is landslides. A tsunami may be generated by a landslide that starts out above sea

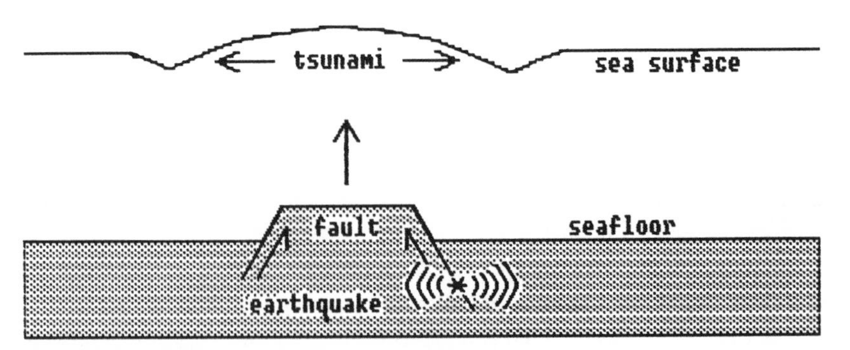

Figure 2.2 Graphic representation of faulting at the ocean floor, producing an earthquake and tsunami.

level then plunges into the sea, or by a landslide that occurs entirely underwater (Figure 2.3).

The highest tsunami waves ever reported were produced by a landslide at Lituya Bay, a confined fjord in Alaska. Here on July 9, 1958, a massive rockslide at the head of the bay produced a tsunami wave that surged up to a high-water mark more than 1,740 feet above the shoreline! The wave then swept out of the bay carrying ahead of it a fishing boat with two people aboard. They miraculously survived and estimated that they had cleared the spit across the mouth of the bay by more than 100 feet!

A small landslide-generated tsunami occurred on the island of Hawaii on August 21, 1951, when a large section of the high cliff overhanging Kealakekua Bay plunged into the water. Small waves, some 2 feet high, crossed the bay splashing ashore at the village of Napoopoo.

The third cause of tsunamis is volcanic activity. Numerous tsunamis have been reported to be the result of nearshore or underwater volcanoes. In most cases, a flank of the volcano is suddenly uplifted or depressed, producing a tsunami in much the same way as those produced by submarine faulting activity. Tsunamis have been produced, however, by the actual explosion of submarine or shoreline volcanoes (Figure 2.4). In 1883, the violent explosion of the famous island volcano, Krakatoa, sent tsunami waves as high as 130 feet crashing ashore in Java and Sumatra. In all, over 36,000 people were killed as a result of the tsunami waves from Krakatoa.

Although tsunamis caused by landslides or volcanic activity may be very large and cause great damage near their sources, they have rela-

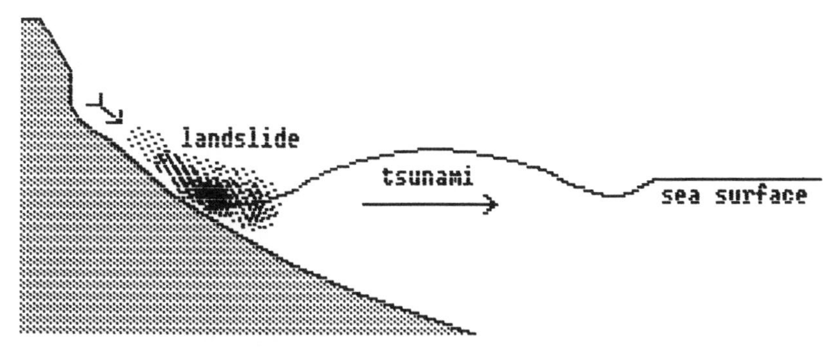

Figure 2.3 Graphic representation of a tsunami generated by landslide originating above sea level.

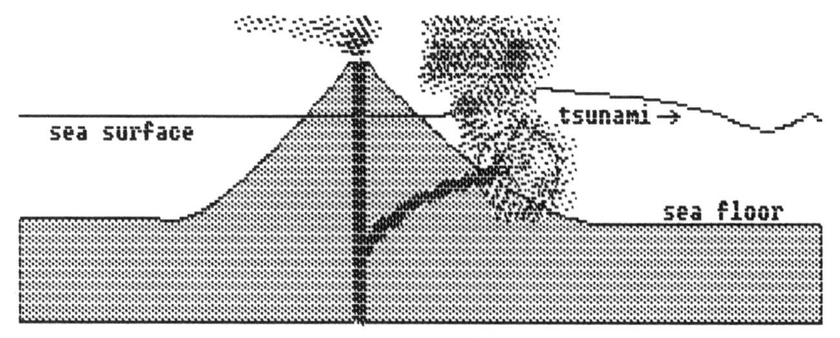

Figure 2.4 Graphic representation of a tsunami produced by submarine volcanic explosion.

tively little energy. They decrease rapidly in size, becoming small or even unnoticeable at any great distance. The giant tsunami waves that cross entire oceans are almost all caused by submarine faulting associated with large earthquakes.

Most tsunamis occur in the Pacific Ocean. This is because the Pacific Ocean basin is surrounded by a zone of very active features in the earth's crust—deep ocean trenches, explosive volcanic islands, and dynamic mountain ranges. Frequent earthquakes and volcanic eruptions make the rim of the Pacific basin the most geologically active region on earth.

Tsunami Waves at Sea

Tsunami waves are very different from other ocean waves. Ordinary waves, which are caused by the wind blowing over the water, are rarely longer than 1,000 feet from crest to crest. Tsunami waves, on the other hand, often measure 100 miles from crest to crest! But perhaps the most astonishing aspect of tsunami waves is their speed. Ordinary wind-generated waves never travel more than 60 miles per hour and are usually much slower. In the deep waters of an ocean basin, tsunami waves may travel as fast as a jetliner—an astonishing 500 miles per hour! Yet these incredibly fast waves may be only a foot or two high in deep water. For these reasons, tsunami waves pass completely unnoticed by ships at sea. In fact, the master of a ship lying offshore near Hilo during the 1946 tsunami stated that he could feel no unusual waves, although he could see great waves breaking on shore.

A popular misconception is that there is only one giant wave in a tsunami. On the contrary, a tsunami may consist of ten or more waves forming what is called a "tsunami wave train." The individual waves follow one behind the other, between 5 and 90 minutes apart.

Tsunami Waves Nearshore

As the tsunami waves move into shallower water and begin to approach shore, they start to change. The shape of the nearshore seafloor, or local submarine topography, has an extreme effect on how the tsunami waves behave. Tsunami waves tend to be smaller on small isolated islands, such as Midway or Wake, where the bottom drops away quickly into deep water. On large islands, such as the main Hawaiian Islands, the tsunami waves have sufficient time to feel bottom and, so to speak, gain a foothold.

As the waves move into the island chain, they begin to be strongly influenced by the sea bottom and are bent around islands and can even be reflected off the shoreline. The reflected waves may interfere with other waves and create extremely large wave heights in unexpected places.

As the wave forms continue to approach shore, they travel progressively more slowly, with the forward speed dropping to around 40 miles per hour. At the same time, the speed of the water particles themselves increases to around 5 to 6 miles per hour. At this point the wave height usually begins to increase dramatically. A tsunami wave that was 2 feet high at sea may become a 30-foot giant at the shoreline.

It is commonly believed that the water recedes before the first wave of a tsunami crashes ashore. In fact, the first sign of a tsunami is just as likely to be a rise in the water level. Whether the water first rises or falls depends on what part of the tsunami wave train first reaches shore: a wave crest causes a rise in the water level and a wave trough causes a recession of the water. Most observers of the 1946 tsunami reported that the first indication of the waves was a withdrawal of the water. However, instruments in both Honolulu and Waimea, Kauai, recorded an initial small rise in the water level. The small rise was probably easily overlooked by unprepared observers, whereas the major withdrawal that followed did not escape notice.

Like storm waves, tsunami waves are often very high on headlands, where the wave energy is concentrated. Unlike ordinary waves, however, tsunamis are also often quite large in bays. In this way tsunami waves, owing to their long wavelengths, do resemble tides and may become amplified in long, funnel-shaped bays. In fact, some of the greatest wave heights ever observed have occurred in such bays.

Another wave phenomenon may also be produced in bays when a tsunami strikes. The water in any basin, be it a bathtub or an ocean basin, will tend to slosh back and forth in a certain period of time determined by the physical size and shape of the basin. This sloshing is known as a seiche. A tsunami wave can set off a seiche, and, if the water motion of the seiche acts in concert with a subsequent tsunami wave, it may cause the water to reach even greater heights than if the tsunami wave acted alone. Much of the great height of tsunami waves in bays may be explained by this combination of a seiche wave and a tsunami wave arriving at the same time. Once the water in a bay is set in motion, the seiche may continue for several days after the tsunami waves have passed.

The presence of a well-developed fringing coral reef off a shoreline also appears to have a strong effect on tsunami waves. A reef may serve to absorb a significant amount of the wave energy, reducing the height and intensity of the wave impact on the shoreline itself.

The 1946 tsunami was less severe along the reef-protected northern coast of Oahu than along the unprotected northern coast of the island of Hawaii. The most extensively developed coral reef system in the Hawaiian Islands lies inside Kaneohe Bay on the east shore of Oahu.

In Kaneohe Bay the 1946 tsunami waves were no more than 2 feet high, whereas just outside the bay on the end of the Mokapu Peninsula, wave heights reached more than 20 feet.

The Bore Phenomenon

The popular image of a tsunami wave approaching shore is that of a nearly vertical wall of water, similar to the front of a wave breaking in the surf. In actuality, most tsunamis probably do not form such wave fronts; instead the water surface is very close to horizontal and it moves up and down. Under certain circumstances, however, an arriving tsunami wave can develop an abrupt, steep front, which will move inland at high speeds. This phenomenon, generally encountered under other circumstances only as a tidal phenomenon, is known as a bore.

Bores produced by tides occasionally occur in the mouths of rivers; some of these have been studied intensively. Well-known examples occur on the Solway Firth and the Severn River in Great Britain, on the Petitcodiac River in Maine, near the mouth of the Amazon River in Brazil, and most strikingly on the Qiantang River in China where the bore may attain a height of 15 feet. In place of the usual gradual rise of the tide, the onset of high tide is delayed; however, when it does occur it takes place very quickly, with the rapidly moving wall of water being followed by a less steep, but still quite dramatic, rise in the water level, accompanied by swift upstream currents. In the Qiantang, current speeds in excess of 10 miles per hour are known.

Similar phenomena have been observed during tsunamis. During the 1946 tsunami in Hilo Bay, a bore estimated to be between 6 and 8 feet high was photographed as it advanced up the Wailuku River (Figure 2.5). The photograph shows the wave front of the tsunami bore travelling over the relatively undisturbed water of the river beneath it. As far as can now be determined, the bore was of limited extent, restricted to the mouth of the Wailuku River where conditions were favorable for its formation.

Under certain conditions, a much larger and more widespread bore may be formed. A study of the 1960 tsunami found that the third wave to enter Hilo Bay developed a bore that may have reached a height of 35 feet. In Chapter 5, the relationship of wave speed to

Figure 2.5 Bore advancing past the railroad bridge at mouth of the Wailuku River, Hilo Bay, during the 1946 tsunami.

water depth, which can cause such a bore to be formed, will be examined. It is clear that such conditions are, fortunately, rather rare.

Although some tsunami waves do indeed form bores, more often the waves arrive like a very rapidly rising tide that just keeps coming and coming. In the case of the 1946 tsunami, the water rose gently at some places, flooding over coastal lands with no steep wave front developing. This is how Bunji Fujimoto remembers the waves at Laupahoehoe. In most areas, however, the advance of the water was accompanied by great turbulence and loud roaring and hissing noises. The normal wind waves and swells may actually ride on top of the tsunami, causing yet more turbulence and bringing the water level to even greater heights.

Because the height of tsunami waves is strongly influenced by the submarine topography and shape of the shoreline, as well as by reflected waves, and because it may be further modified by seiches, tides, and wind waves, the actual inundation and flooding produced by a tsunami may vary greatly from place to place over only a short distance. The Aleutian tsunami of 1946 produced waves 30 feet high

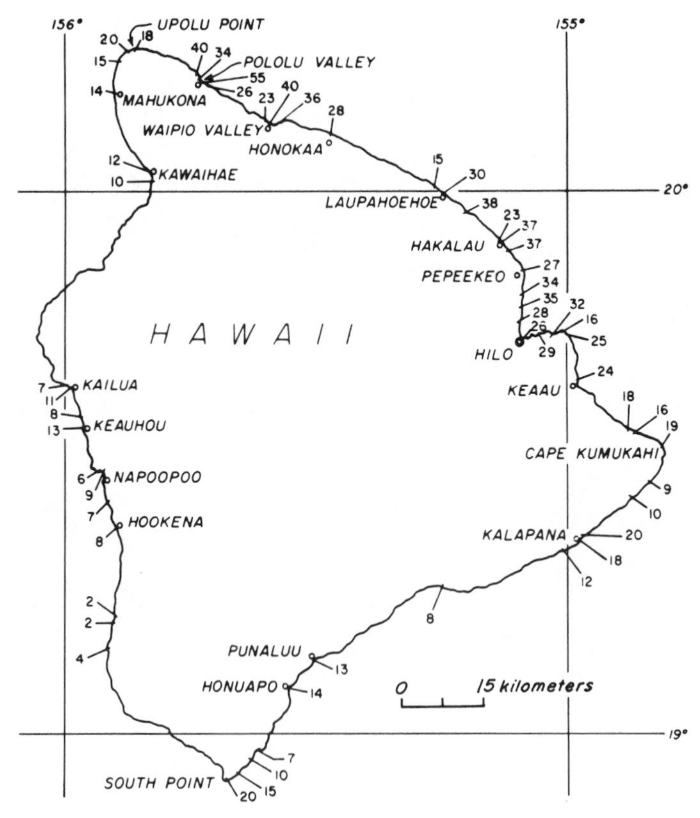

Figure 2.6 Map of the island of Hawaii, showing heights (in feet above mean lower low water) reached by the water during the tsunami of April 1, 1946.

at Laupahoehoe, while just a few miles farther up the coast the waves were only half as high (Figure 2.6).

Just as the wave height and inundation can vary greatly from place to place, so too can the largest wave in the tsunami wave train. Reports of the 1946 tsunami generally agree that the third or fourth waves were the highest and most violent; however, at the Waimea River on Kauai, the sixth crest was the highest and most destructive.

Wave Damage and Destruction

The tremendous destruction from tsunamis such as that which occurred in Hilo in 1946 is produced in several ways. The sheer force of the moving wall of water in a bore literally can raze almost every-

Figure 2.7 Flood damage to downtown Hilo caused by the 1946 tsunami.

thing in its path. It has been estimated that the force of the water in a bore could momentarily attain the enormous pressure of 2,000 pounds per square foot.

Although the most dramatic image associated with a tsunami is the bore, it is the flooding effect of a tsunami that causes the most damage. This was the case in Hawaii in 1946, and is well illustrated in photographs of the damage to downtown Hilo (Figure 2.7). Two different terms are often used to describe the extent of tsunami flooding (Adams 1985): inundation and run-up. Inundation is the depth of water above the normal level, and is usually measured from sea level at average low tide. Inundation may be measured at any location reached by the tsunami waves. Run-up, on the other hand, is the inundation at the maximum distance inland from the shoreline reached by the tsunami waters.

Even the withdrawal of the tsunami waves can cause significant damage. As the water is rapidly drawn back toward the sea, it may scour out bottom sediments, undermining the foundations of buildings. Entire beaches have been known to disappear as the sand is carried out to sea by the withdrawing tsunami waves. During the 1946 tsunami, the outflow of the water was rapid and turbulent, making loud hissing, roaring, and rattling noises. At several places, houses were carried out to sea with the receding water.

The advance and retreat of tsunami waves causes the water level in ports, harbors, channels, and other navigable waterways to change radically, creating treacherous and unpredictable currents. Boats of all sizes are ripped from their moorings, smashed together, tossed ashore, sunk, or carried out to sea.

Defending Against the Waves

Can anything be done to prevent the damage and loss of life from destructive tsunamis?

After the 1946 tsunami it was noted that the breakwaters at Hilo and at Kahului on Maui helped reduce the impact of the waves in their respective harbors. It was suggested that perhaps breakwaters or seawalls could be built as a defense against tsunami waves. But although the breakwaters did reduce the waves, most experts agreed that breakwaters or seawalls could not be built high enough or strong enough to hold the water back completely.

Can waterfront buildings be built to withstand the force of the tsunami waves? After the 1946 tsunami it was observed that houses elevated on stilts survived the waves much better than those built directly on the ground. Apparently the water was able to pass under such houses without greatly disturbing them. Reinforced concrete structures (like the Coca Cola bottling plant) were least affected by the waves: concrete structural supports were often left standing while weaker non-structural walls were carried away.

It is no accident that the hotels in Hilo that now stand along Banyan Drive facing Hilo Bay have open ground floors with high ceilings. This particular feature is not purely for aesthetics. The tsunami-resistant design concept developed for major buildings such as hotels is to have the first-floor living area elevated above the potential wave height and to assume that the ground floor and basement will be inundated. The structural walls and columns on the ground floor are designed to resist the impact forces of the waves, while the nonstructural walls between the columns are designed to be expendable as the waves pass through the building.

Following the disastrous tsunami of 1946, much of the downtown bayfront area of Hilo between the Wailuku and Wailoa rivers was left as a parkway. It was hoped that this barrier would serve as a buffer against the destructive action of future tsunamis.

But what of the human loss, the tragic deaths caused by the totally unexpected assault of a tsunami? Could anything be done to warn the population of the approach of the waves?

CHAPTER THREE

The Tsunami Warning System

FOLLOWING the tragic loss of life in Hawaii as a result of the 1946 Aleutian tsunami, the population wanted to know if anything could be done to warn of the approach of these catastrophic waves. But the newspaper headlines read "Warning Impossible, Geodetic Chief Asserts," and the Commerce Department denied that its Coast and Geodetic Survey was remiss in not warning the population. Yet, oddly enough, warnings of tsunamis had been issued in Hawaii during the 1920s and 1930s by the Hawaiian Volcano Observatory. Why had Hawaii been caught completely unprepared in 1946, when warnings were possible more than 20 years earlier? To answer this question, we need to take a look at the early records of tsunamis in Hawaii.

During the nineteenth century, numerous tsunamis (referred to as tidal waves) were reported in newspapers and magazines in Hawaii. From reading these accounts, it isn't always possible to know whether they refer to a genuine tsunami or very large, wind-generated storm waves. Also, because news travelled slowly in those days, it might take months for knowledge of an earthquake in Alaska or South America to become known in Hawaii. Dates were sometimes mixed up and, as a consequence, the fundamental relationship between earthquakes and tsunamis remained obscure.

Finally, toward the end of the nineteenth century, a seismological station, which could record even distant earthquakes, was established in Honolulu. It became possible to associate a large tsunami in Hawaii with an earthquake in, for example, South America.

In 1912, the Hawaiian Volcano Observatory (HVO) was established: on its staff were a number of scientists interested in the study of earthquakes and tsunamis. Thomas A. Jaggar, the founder of HVO and its director until 1940, began to investigate and report on tsunamis, including research on historical accounts of earlier tsunami events.

Jaggar knew that the seismic waves caused by earthquakes are transmitted across the globe in a matter of minutes. A large earthquake in Chile, for example, would be registered on seismographs in Hawaii hours before a tsunami could reach the islands. Why not use this lead time to warn of an impending tsunami?

In early 1923, Jaggar had a firsthand opportunity to witness the earthquake-tsunami relationship. As he inspected the seismograph at 8 A.M. on February 23, he immediately noticed the trace of a large earthquake that had been recorded earlier at 5:32 A.M. local time. Jaggar quickly calculated that the epicenter would be about 2,500 miles away, possibly under the sea off the Aleutian Islands. The seismic waves had taken only about 7 minutes to travel through the earth to Hawaii from the Aleutians; if tsunami waves were on the way, they would arrive several hours later. Jaggar notified the harbormaster of the possibility of a tsunami later that day, but his warning was not taken seriously.

The first waves began to arrive at Haleiwa on the north shore of Oahu at 12:02 P.M. The waves would arrive later on Maui, where they caused serious damage at Kahului. At 12:30 P.M. the tsunami struck Hilo, almost exactly 7 hours after the earthquake occurred in the Aleutians. The largest wave was the third of the series, rising to more than 20 feet at Waiakea. Surging into Hilo Bay, the rising sea carried the local fishing fleet of sampans from their moorage in the Wailoa River and smashed the vessels into the railroad bridge. The railway embankment and bridge were destroyed and one man was killed.

It was a painful lesson, but it did point out the possibility of protecting life and property by warning of the approach of tsunami waves. Later that same year a meteorologist, R. H. Finch, gave a speech at a scientific meeting in Sydney, Australia. The title of his talk was "On the Prediction of Tidal Waves," a subject, which as Jaggar prophetically stated, "might well be studied to advantage."

Finch pointed out that the time in hours it took tsunami waves to reach Hawaii was approximately equal to the time in minutes it took the seismic waves to travel the same distance. Using the minutes (for earthquake waves) = hours (for tsunami waves) rule, he felt that it should be possible to predict the arrival time of tsunami waves from all parts of the Pacific. Finch suggested also that since most seismographs were inspected rather infrequently, some type of "alarm bell" could be attached to the instruments to alert scientists when a large earthquake was registered.

Jaggar felt that the study of tsunamis was now more important than ever. He urged that to predict accurately the time it took the waves to travel from their sources to Hawaii, it would be necessary to correlate actual earthquakes with the arrival times of tsunami waves. He also thought that more intensive study of earthquakes and tsunamis would make it possible to determine the earthquake intensity required to generate a tsunami.

Fortunately, most tsunamis are very small and go unnoticed unless registered on a tide gauge. In order to benefit from the study of these small tsunamis, Jaggar travelled to Washington, D.C. to arrange with the U.S. Coast and Geodetic Survey for the establishment of a tide gauge at Hilo.

With tide gauges established in Honolulu and Hilo it was possible to examine the records of changes in water level and to identify even small tsunami waves of a foot or less. With seismographs, tide gauges, and the research of knowledgeable scientists in Hawaii, the connection between earthquakes and tsunamis became better known.

Another opportunity to test the earthquake-tsunami relationship came in 1933. At 7:10 A.M. Hawaii time on March 2, a large earthquake was registered on the seismographs at HVO and on a similar instrument in Kona operated for the Hawaiian Volcano Research Association by Captain R. V. Woods. Woods was a retired sea captain who, as a volunteer, inspected the Kona seismograph daily and sent recordings by mail to HVO weekly. Reading the seismic waves, he calculated the distance to the earthquake epicenter as 3,950 miles—possibly off the coast of Japan.

Meanwhile at HVO, seismologist A. E. Jones also concluded that a large earthquake had occurred off the coast of Japan. Knowing that a tsunami might possibly have been generated by the earthquake, Jones notified the harbormaster at 10 A.M. and indicated that waves might begin to arrive about 3:30 that afternoon.

In Kona, Captain Woods notified the Captain Cook Coffee Company at Napoopoo that tsunami waves might arrive at about 3 P.M. Remembering the 1923 tsunami, cargo was removed from the dock at Napoopoo and the Hilo sampan fleet moved out to anchorages in the harbor.

About noon, news broadcasts on the radio announced that a disastrous earthquake had occurred in Japan. In fact, the earthquake and accompanying tsunami in Japan resulted in nearly 1,600 deaths and more than 2,800 homes being washed away. The tsunami waves were highest along the northeast coast of the main Japanese island,

Honshū, centered around Ryori Bay. Waves up to 96 feet high were reported at the head of the funnel-shaped bay. Tsunami waves were also spreading across the Pacific at nearly 500 miles per hour.

On the Kona side of the island of Hawaii—facing Japan—the first waves of the tsunami arrived at 3:20 P.M. local time. At first the sea withdrew, exposing wide areas of the sea bottom in the bays at Kailua, Keauhou, Napoopoo, and even at Kaalualu near South Point. Canoes and other small craft in Kailua Bay were torn from their moorings and capsized. As the sea returned, walls were washed down, houses were flooded and moved, and, in Kailua, a sampan on the marine railway was tossed over the seawall—landing relatively undamaged on the other side. Of the series of some ten waves, the last was the most damaging. At Napoopoo the water dropped 8 feet below mean tide and then rose again 9 and a half feet, for a total vertical range of 17 and a half feet.

In Hilo the waves began to arrive at 3:36 P.M., but the water only rose and fell a total of 3 feet and caused no property damage.

Thanks to the warnings of Jones in Hilo and Woods in Kona there were no lives lost in Hawaii during the 1933 tsunami. Yet these tsunami warnings based on the occurrence of earthquakes led to false alarms. Soon the population began to disregard the warnings and it was realized that a tsunami warning system based solely on the occurrence of submarine earthquakes was virtually useless. Immediately following the 1946 tsunami, in an article in the *Honolulu Advertiser* (2 April 1946), the commander of the U.S. Coast and Geodetic Survey stated the case: "Less than one in one hundred earthquakes result in tidal waves and you don't alert every port in the Pacific each time a quake occurs."

We now know for a fact that only a small number of undersea earthquakes are accompanied by tsunamis. This is probably because most movement of faults takes place below the surface of the earth's crust and causes no movement of the seafloor itself which would displace water. Also, it takes seafloor motion producing a fairly large earthquake to generate a tsunami. Records of past tsunamis and earthquakes show that a quake of at least Richter* magnitude 7 is required to cause a dangerous tsunami in Hawaii. In general, it

---

*The most commonly used scale for measuring the magnitude of earthquakes was devised by C. F. Richter and is known as the "Richter scale." The scale is not linear but logarithmic, so that each unit represents a 10-fold increase in ground movement and a 32-fold increase in energy. For example, an earthquake of Richter magnitude 7 would have 10 times the earth shaking and release 32 times as much energy as an earthquake of magnitude 6.

appears that the larger the earthquake the larger the tsunami it can generate, although this is not always strictly true.

In spite of government claims that no warning of the 1946 tsunami was possible, it was obvious that something had to be done to protect the population of Hawaii. Both civilian and military sources criticized the U.S. Coast and Geodetic Survey for not issuing a warning. After all, as critics pointed out, the seismic waves from the earthquake had been recorded at HVO within minutes after the earthquake struck the Aleutians; consequently, a tsunami could have been predicted.

By coincidence, in 1946 a number of scientists happened to be in Hawaii in connection with the Bikini atomic bomb tests. Because these scientists observed the tsunami firsthand, it became the most thoroughly studied tsunami in history. In a scientific paper published in 1947, Gordon Macdonald, Francis Shepard, and Doak Cox stated that they felt the loss of life from tsunamis could be largely avoided (Macdonald et al. 1947). They recommended that a system of stations, which would observe the arrival of the large, long waves of tsunamis, be established around the shores of the Pacific and on mid-Pacific islands. The arrival of these waves would then be reported immediately to a central station, whose duty would be to correlate the reports and issue warnings to places in the path of the waves. They felt that such a system would make it possible to give people in the Hawaiian Islands enough warning of the approach of a tsunami to permit them to reach places of safety. Their ideas combined the advance indication of the "possibility" of the generation of a tsunami, as indicated by the measurement of a large earthquake, with the "confirmation" of a tsunami through measurements from tide gauges lying along the path of the waves. Such a system could go a long way toward eliminating the problem of false alarms.

A necessity for a workable warning system was to develop a method for quickly and accurately determining the travel-time for tsunami waves to arrive in Hawaii from various earthquake-producing areas around the Pacific. This problem was solved in early 1947 by the preparation of a tsunami travel-time chart for Honolulu.

Another obstacle to a workable warning system lay with the seismograph instruments then in use. Visible earthquake-recording systems in existence in 1946 were of poor accuracy, and the more accurate film-recording instruments were read only once daily when the film was developed. Various new devices were tried out, and in 1947 and 1948 equipment designed by Fred Keller, a scientist living in

Pennsylvania, was built and installed at Tucson, Arizona; College, Alaska; and Honolulu, Hawaii. With these new instruments, highly accurate seismograph records were continuously available for inspection, and when a strong earthquake was registered, an alarm was sounded.

Finally, it was essential to establish a rapid, high-priority communications system to transmit reports on the earthquakes and tsunami waves. On August 12, 1948, a tentative plan was approved.

An official tsunami warning system was established by the U.S. Coast and Geodetic Survey and called the Seismic Sea Wave Warning System; a name which was later changed to the Tsunami Warning System (TWS). The system comprised the Coast and Geodetic Survey seismograph observatories at College and Sitka, Alaska; Tucson, Arizona; and Honolulu, Hawaii; and tide stations at Attu, Adak, Dutch Harbor, and Sitka, Alaska; Palmyra Island; Midway Island; Johnston Atoll; and Hilo and Honolulu, Hawaii. The Honolulu (seismic) Observatory, located at Ewa Beach on the island of Oahu, was made the headquarters (Figure 3.1).

Initially, the warning system was to supply tsunami warning information to civil authorities in the Hawaiian Islands and to the various military headquarters in Hawaii for dissemination throughout the Pacific to military bases and to authorities in the islands of the U.S. Trust Territory of the Pacific.

Figure 3.1 Honolulu Observatory, headquarters of the Tsunami Warning System, located at Ewa Beach, Oahu.

Figure 3.2 Row of seismographs at the Honolulu Observatory of the Tsunami Warning System.

How the Tsunami Warning System Works

The tsunami warning system takes full advantage of the relationship between tsunamis and earthquakes. The vast majority of Pacific-wide tsunamis are caused by severe faulting on the ocean floor, and an earthquake of magnitude 7 or greater almost always accompanies the generation of a major tsunami. When a major earthquake does occur, it is recorded by seismographs all over the world within a matter of minutes. Seismograph stations can not only estimate the size of an earthquake, but with reports from three or more stations, the epicenter, that is, the position on the earth's surface directly over the site of origin of an earthquake, can be determined.

The functioning of the Tsunami Warning System begins with the detection of an earthquake by the network of seismograph stations (Figure 3.2). Special seismic alarms are set to go off when an earthquake of magnitude 6.5 or greater occurs anywhere in the Pacific. Scientists at the observatories then rush to their instruments and begin interpreting the seismograms. Their readings are then immediately sent to the Honolulu Observatory. Within a half hour the Honolulu Observatory has determined the epicenter and magnitude of the earthquake, and sent messages to other seismograph stations requesting additional data.

Tsunami Watch

If the earthquake is strong enough to cause a tsunami and if the epicenter is located close enough to the ocean, a Tsunami Watch is declared. A Tsunami Watch is automatically issued for all earthquakes greater than Richter magnitude 7 occurring in the area of the Aleutians and for all quakes greater than 7.5 occurring elsewhere in the Pacific basin.

Agencies such as state and county Civil Defense, police departments, the American Red Cross, and others are alerted that a Tsunami Watch is in progress. Local broadcast media announce the Tsunami Watch over the airwaves.

Now it is time to confirm if a tsunami has, indeed, been generated and if so, how big it is. The first positive indication that a tsunami has been generated comes usually from tide-gauging stations nearest the disturbance. Tsunamis appear on the tide gauge records as distinct abnormalities in the normal curve of the rise and fall of the tides.

A Pacific-wide network of communication channels sends out messages requesting the five nearest tide stations to the epicenter of the earthquake to monitor their gauges. Trained observers at locations in the path of the tsunami are requested to report to the warning center on wave activity in their areas.

If the tide stations report negligible waves or no tsunami, then the Tsunami Watch is cancelled. But if the warning center receives reports from tide gauges or observers indicating that a destructive tsunami posing a threat to the population has been generated, then a Tsunami Warning is issued.

Tsunami Warning

When a Tsunami Warning is declared, the public is informed through the Hawaii State Emergency Broadcast System (EBS), which broadcasts the warning on all commercial radio stations. State and county Civil Defense agencies implement prearranged plans to evacuate the population from low-lying coastal areas that might be threatened by the tsunami.

Areas of the islands likely to be subject to the greatest danger have been predesignated, based on studies of the inundation of previous destructive tsunamis. Evacuation routes for some areas have likewise been preplanned by Civil Defense agencies. Maps of areas in danger

Figure 3.3 Road sign indicating tsunami evacuation route in Hilo.

Figure 3.4 *(opposite)* Map showing travel times to Honolulu (in hours) of tsunamis originating around the Pacific Ocean rim.

of tsunami inundation are printed in the first pages of each island's telephone directory. In the Hilo area, evacuation routes are marked by special road signs so as to try to minimize confusion during an evacuation (Figure 3.3).

Previously, we stated that vessels offshore may be completely unaffected by tsunami waves if they are in deep, open water. In fact, U.S. Coast Guard vessels always put out to sea upon being advised of a Tsunami Warning. Once the dangerous waves have passed, the Coast Guard vessels then return inshore to begin rescue operations. Private boat owners are also warned to put their vessels out to sea upon the announcement of a Tsunami Warning. In many cases, preparing vessels for putting out to sea may need to begin earlier—during the Tsunami Watch. Boat owners are further advised that local sea conditions nearshore may remain dangerous and unpredictable for several hours after the last major tsunami wave, and may even take several days to return to normal.

Wave Arrival Time

One of the most important aspects of the entire Tsunami Warning System is its ability to predict the actual arrival time of the first tsunami waves. This capability is based on one of the special characteristics of tsunami waves—their great wavelength. In order to understand just how the arrival time of a tsunami is determined, we need to know a little more about waves in general.

Oceanographers divide waves into various categories based on their wavelength and the depth of water through which they pass. When the water depth is less than 1/20 the wavelength, the waves are

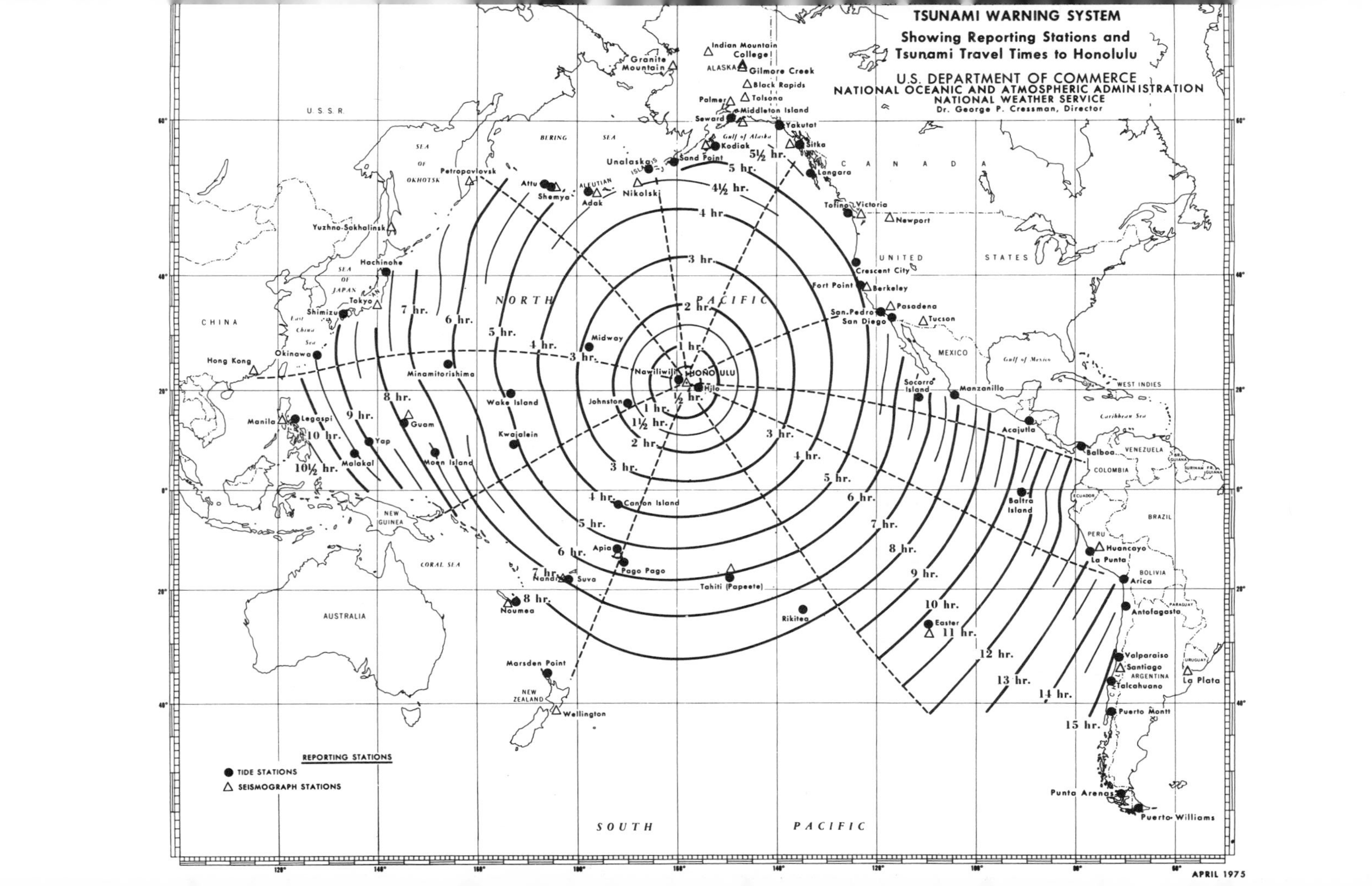
TSUNAMI WARNING SYSTEM
Showing Reporting Stations and
Tsunami Travel Times to Honolulu
U.S. DEPARTMENT OF COMMERCE
NATIONAL OCEANIC AND ATMOSPHERIC ADMINISTRATION
NATIONAL WEATHER SERVICE
Dr. George P. Cressman, Director
REPORTING STATIONS
TIDE STATIONS
SEISMOGRAPH STATIONS
APRIL 1975
NORTH PACIFIC
SOUTH PACIFIC
U. S. S. R.
CHINA
CANADA
UNITED STATES
MEXICO
AUSTRALIA
NEW ZEALAND
NEW GUINEA
COLOMBIA
VENEZUELA
BRAZIL
PERU
BOLIVIA
ARGENTINA
ALASKA
BERING SEA
SEA OF OKHOTSK
SEA OF JAPAN
CORAL SEA
WEST INDIES
HONOLULU
Hilo
Nawiliwili
Midway
Johnston
Wake Island
Kwajalein
Minamitorishima
Guam
Yap
Malakal
Moen Island
Canton Island
Apia
Pago Pago
Suva
Nandi
Noumea
Tahiti (Papeete)
Rikitea
Easter
Baltra Island
Marsden Point
Wellington
Okinawa
Hong Kong
Manila
Legaspi
Shimizu
Tokyo
Hachinohe
Yuzhno-Sakhalinsk
Petropavlovsk
Attu
Shemya
Adak
Nikolski
Unalaska
Sand Point
Kodiak
Seward
Palmer
Middleton Island
Tolsona
Black Rapids
Gilmore Creek
Indian Mountain College
Granite Mountain
Yakutat
Sitka
Langara
Tofino
Victoria
Newport
Crescent City
Fort Point
Berkeley
San Pedro
San Diego
Pasadena
Tucson
Socorro Island
Manzanillo
Acajutla
Balboa
La Punta
Huancayo
Arica
Antofagasta
Valparaiso
Santiago
Talcahuano
La Plata
Puerto Montt
Punta Arenas
Puerto Williams
½ hr.
1 hr.
1½ hr.
2 hr.
3 hr.
4 hr.
4½ hr.
5 hr.
5½ hr.
6 hr.
7 hr.
8 hr.
9 hr.
10 hr.
10½ hr.
11 hr.
12 hr.
13 hr.
14 hr.
15 hr.

known as shallow-water waves. For example, a wave with a wavelength of 20 feet would be considered a shallow-water wave when passing through water less than 1 foot deep. The significance of this is that the speed of a shallow-water wave is determined solely by the depth of water. In other words, if we know the water depth, we can calculate the velocity of any shallow-water wave.

But how can tsunamis possibly be considered shallow-water waves when they cross the deep ocean? Remember that tsunami waves may have a wavelength of more than 100 miles. If the water depth is less than 1/20 of the 100-mile wavelength (or 5 miles) then the tsunami waves would be considered shallow-water waves. Most of the deep Pacific Ocean basin is less than 3 miles deep, so tsunamis are, indeed, shallow-water waves. As a result, in order to determine their speed, all we need to know is the depth of the water through which they pass.* The depths of most of the ocean basins have been charted—at least on a gross scale—therefore, we can calculate the time for a tsunami wave to travel between any two points in the ocean. In most of the deep sea this works out to around 450 miles per hour, but the speed will vary depending on the exact water depth along the path of the tsunami.

A travel-time chart centered on Honolulu is shown in Figure 3.4. Using this chart and knowing the area of origin of a tsunami, it is possible to predict the travel-time for the waves to reach Honolulu. For example, in the case of a tsunami originating in the Aleutians near Adak, Alaska, the chart shows that it would take 3 and a half hours for the waves to reach Midway Island, and 4 and a half hours to reach Honolulu. By the time a Tsunami Watch is declared, the estimated arrival time of the first wave has already been calculated by computer for 62 different sites around the Pacific, and messages have been sent out to all stations. Evacuation procedures are then coordinated around the predicted arrival time of the tsunami.

The decade of the 1950s dawned with high hopes for the newly established Tsunami Warning System. A network of seismic observatories and tide stations was in place. Communication and evacuation procedures had been established. All that was needed to test the system was a Pacific-wide tsunami. The wait would not be long!

---

*The basic equation for determining the speed of a tsunami wave is $C = \sqrt{gh}$ (where "C" is the wave velocity, "g" is the coefficient of gravity, and "h" is the depth of water).

CHAPTER FOUR

The Warning System in Action: The Tsunamis of 1952 and 1957

The Tsunami of 1952

AT 1652 Greenwich mean time (GMT)* on November 4, 1952, a strong submarine earthquake occurred in the far northwest Pacific basin. Fifteen minutes later, at 1707 GMT, the earthquake alarm at the Honolulu Observatory was set off as the shock was registered on seismographs. The Tsunami Warning System immediately went into action. The seismic observatories in California, Arizona, and Alaska all reported additional information on the earthquake. In less than an hour, the Warning System had determined the earthquake's epicenter to be 51° north latitude, 158° east longitude, off the southeastern coast of the Kamchatka Peninsula of the Soviet Union.

A Tsunami Watch was issued. The Warning System calculated that the first waves would arrive at Honolulu at 2330 GMT (1:30 P.M. Hawaii time). The police and military authorities were notified and constantly updated by the Warning System as reports of the progress of the tsunami waves came in from points closer to its source.

Midway Island, several hundred miles closer to the origin of the tsunami, was struck before the main Hawaiian Islands. Midway reported that it was covered with 3 feet of water.

The tsunami reached Hilo just after 1:30 P.M. The tide gauge at the end of Pier 1 recorded a rise in the water level of 4 feet at 2332 GMT, with no sign of an initial withdrawal prior to the rise. The water continued to rise until it reached a height of approximately 7

*Greenwich mean time (GMT) is the time at the prime meridian, i.e., 0° longitude, which passes through the Royal Observatory in Greenwich, England. Scientists use Greenwich mean time for events such as earthquakes and tsunamis. Hawaii time is ten hours behind Greenwich time, i.e., GMT − 10. For example, when it is noon at Greenwich, England, it is only 2 A.M. in Honolulu.

feet above "mean lower low water" (MLLW)* at 2349. By 0005 GMT (November 5), 16 minutes later, the water level had fallen to about 1 foot below MLLW—the tide gauge showing a total change in water level of some 8 feet. The tide gauge, dampened against very rapid changes in water level, may not have registered the full extent of the inundation. Marks on the sides of the wharf shed on the Hilo commercial pier indicate that the wharf deck was flooded to a depth of nearly 2 feet, some 11 and a half feet above MLLW.

As the water flooded across the pier, freight awaiting shipment was washed into the bay. A new boat landing for the harbor pilot, near Pier 2, was lifted from its foundation by the first wave and totally demolished as it was dropped by the following trough.

As the tsunami moved into Hilo Bay it swept over Cocoanut Island, 12 feet above MLLW. The walls of buildings on the island were literally punched out by the force of the water. The small bridge connecting Cocoanut Island with the shore was floated off its foundation and smashed as it was dropped by the following trough.

At Reeds Bay in Hilo the water rose as much as 11 feet, destroying a house as the supports were washed out and carrying a fishing sampan 200 feet inland.

The mouths of both the Wailuku and Wailoa rivers were flooded to 9 feet above normal, and a small 18-inch bore surged into the Wailoa Estuary during one of the later waves.

The first large waves were followed by a series of ever-diminishing smaller waves with an average period between waves of about 18 minutes. These smaller waves continued for the next several days.

On the island of Hawaii only the Hilo area received significant damage. Although the waves were too small to measure along many parts of the coast, Hilo Bay would record the largest waves in the Hawaiian Islands. Even in nearby Keaukaha the rise in water was so small as to go unnoticed by many. Hilo had proved to be vulnerable not only to Aleutian tsunamis but also to those from the coast of northeast Asia.

Buildings along Kamehameha Avenue in Hilo were protected from flooding by the embankment of the new coastal highway, but damage to those buildings lying along the shoreline between the Wailoa

---

*The reference sea level used in measuring the depths of water on marine charts and tide calendars in most parts of the Pacific Ocean is "mean lower low water" (MLLW). This is the average of the lowest low tides in a particular area.

River and the breakwater was extensive. Several fishing sampans were beached, and others in a nearby boatyard were damaged.

The waves of the 1952 tsunami were not nearly as high as those of 1946. Although lawns were flooded, telephone lines knocked down, automobiles marooned, and a cement barge was hurled against a freighter in Honolulu Harbor, the death toll was limited to six cows! Not a single human life was lost, and property damage was estimated at less than $800,000.

The Warning System had worked and there can be no doubt that it saved lives. However, one disturbing event did occur during the 1952 tsunami. Sources not connected with the Warning System issued statements that the first or second wave crests were the dangerous ones, and that thereafter the danger was over and people in threatened coastal areas could safely return to their homes or places of business. The Warning System, in fact, had received a report from the commander of the ship the *Hawaiian Sea Frontier* warning that the fourth wave had done the most damage at Midway Island, and suggesting caution against "securing" early. Furthermore, as mentioned earlier, studies of the 1946 tsunami showed that at different localities the largest waves ranged from the third to the eighth. The only measure of safety for residents of coastal areas is to stay away from the shore for at least two hours after the waves have stopped coming in.

The Tsunami of 1957

More than ten years had passed since the disastrous Aleutian tsunami of 1946, when on March 9, 1957, at 1422 GMT a major earthquake again shook the seafloor near the Aleutian Trench. It was 4:22 A.M. Alaska time in the fishing village of Unalaska, when observers at the tide station were awakened by the shock. Eight minutes later, at about 4:30 A.M., the seismograph at the Honolulu Observatory registered the quake. The seismic alarms were set off, including the alarm in the nearby home of seismologist Harold Krivoy, who jumped out of bed and rushed to the observatory. Messages were immediately sent out to the observatories in Tucson, Arizona, and Sitka, Alaska, requesting additional information. Reports from the observatories calculated the earthquake's epicenter to be at 51.3° north latitude, 175.8° west longitude, south of the Andreanof Islands. The magnitude measured between 7.75 and 8.3 on the Richter scale; it was definitely an earthquake capable of causing a destructive tsunami. If a tsunami were on the way, it would reach Hawaii between 8 A.M. and 9 A.M. that day. A Tsunami Watch was put into effect.

Coastal communities in the Aleutians were contacted over military communication lines and requested to keep a close watch on the tides. At 6:50 A.M., Unalaska reported that the level of the sea had begun to rise. At 6:53 A.M., Adak reported that waves 8 feet above normal had begun to break along the shoreline. A tsunami was on the way and it was time for a Tsunami Warning.

The Warning System immediately alerted the police and military authorities. The police notified Civil Defense headquarters, the Fire Department, and the Honolulu City-County Emergency Hospital. Meanwhile, the Warning System radioed messages to police and Civil Defense on the neighbor islands. Commercial radio stations began to broadcast the Tsunami Warning.

On all the Hawaiian Islands, police and firemen were clearing the beaches and warning residents in low-lying areas, in addition to alerting people driving along coastal highways. Fire stations and ambulance crews began to prepare for rescue operations.

U.S. Army and Marine helicopters were readied for flight. The U.S. Coast Guard sent messages to harbormasters throughout the islands to warn vessels to put to sea, and both U.S. Navy and Coast Guard vessels themselves began to head offshore.

At 8:44 A.M., the warning center received a message that the first waves had struck Midway Island at 7:45 that morning. In little more than an hour the tsunami would reach the main Hawaiian Islands.

Kauai

Just before 9 A.M. the first waves began to strike Kauai, the most northern of the main Hawaiian Islands. The 1946 tsunami had come ashore on Kauai like a huge surf breaking, but in 1957 the tsunami surged in like a giant flood tide. On the northern coast of Kauai at Haena (facing directly toward the Aleutians), the sea rose more than 32 feet above normal. Of the 29 homes in the small community, only 4 were left standing after the waves.

The villages of Wainiha and Kalihiwai were virtually wiped out as the waves surged ashore on Kauai. Along with Haena, these villages would be isolated for days as the bridges and roads connecting them to the rest of the island were destroyed.

Along the beachfront of quiet, scenic Hanalei bay, many homes were destroyed as the tsunami surged over the reef into the bay and up the normally tranquil river. Norman Kawamoto and Yoshio Miyashiro were sitting in their small boat when the tsunami advanced up the river and capsized them. They clung desperately to their boat as they were washed back and forth by the subsequent waves. Finally, after three hours, they managed to grab hold of some *hau* tree branches, where they stayed until they were rescued by Joseph Nakamura in a small boat. Fortunately, they were to be the only Kauai residents caught by the sea.

In all, between 75 and 80 homes were destroyed or badly damaged on Kauai, more than twice the amount of damage suffered by the island during the 1946 tsunami. Marine Corp helicopter crews worked around the clock to provision the areas of the island isolated by the waves, and the Hawaii National Guard stood on guard against looting.

Oahu and Points South

Not long after the arrival of the waves on Kauai, the tsunami reached the north shore of Oahu. As on Kauai, the tsunami came ashore like a rapidly rising tide that just kept coming and coming. From the polo grounds at Mokuleia to the famous surfing beach at Waimea Bay, the water surged ashore. Farther to the east, past Kahuku Point, the tsunami engulfed beachfront homes at Laie (Figure 4.1).

At Makaha beach on the southwest-facing Waianae Coast, a visitor to the islands was making home movies of the scenic shore. As he filmed his friend standing on a low seacliff the water suddenly rose more than 15 feet, stopping just at the feet of his awe-stricken companion. The waves had begun to bend around the island.

At nearby Pokai Bay more than 50 small boats and half a dozen yachts were smashed against the new breakwater by the surge of the tsunami waves. One boat was washed over 200 yards inland and left to rest in a stand of *kiawe* trees.

In the Ala Wai Boat Harbor, next to Waikiki Beach, strong currents surged in and out—breaking off pilings and docks. Large yachts were washed back and forth dragging sections of pier with them.

Figure 4.1 Oceanfront properties at Laie, Oahu, flooded by the 1957 tsunami.

The tsunami continued to move south, passing Molokai where the taro crop on the eastern end of the island was inundated and ruined. At Kalaupapa, the site of the community for patients of Hansen's disease, the waves rushed ashore to heights of more than 14 feet, smashing the settlement's water pipeline.

At the port of Kahului on Maui, the water surged in and out, creating a powerful vortex from the enormous turbulence (Figure 4.2).

The Island of Hawaii

At 9:17 A.M. the first waves struck Hilo. As the tsunami surged into the bay, the breakwater was inundated. The water rose 10 feet above normal at Pier 1, flooding the wharf by 2 feet and causing extensive damage to cargo in the warehouse. The waves continued to move into the bay, submerging Cocoanut Island by some three feet and, like the tsunamis before it, damaging the small foot bridge connecting the island to the shore.

As the water moved up the Wailoa Estuary, several fishing boats

Figure 4.2 Kahului, Maui, during the 1957 tsunami. Note the whirlpools formed by water withdrawing from the harbor.

were washed ashore or overturned. The high ground around the Wailoa bridge was left as an island in a shallow sea of the tsunami floodwater. Reeds Bay was submerged to a depth of 9 feet, as it had been in both 1946 and 1952.

The pressure of the tsunami surging over the Hilo sewage outfall caused a manhole cover near the base of Mamo Street to be blown off, flooding the street. Buildings were badly damaged along the unprotected sections of the bayfront, but the main area of downtown Hilo was largely spared: the coastal barrier and the parkway between the Wailuku and Wailoa rivers both served as an effective buffer to the waves.

In low-lying Keaukaha the water flooded the ground floors of more than a hundred homes, but caused little major damage. In fact, the total damage to homes, businesses, and boats for the entire Hilo area would amount to a comparatively modest $150,000.

The tsunami waves continued to move south past Hilo to Cape Kumukahi, where logs and cane trash were washed more than 12 feet above MLLW.

The western side of the Big Island, facing away from the source of the tsunami, was the area least affected; however, a 5-to-6-foot wave was reported to have come ashore at Keauhou at about 7:00 P.M., some 10 hours after the first tsunami waves had struck the northeastern side of the island. According to some experts, this delayed wave could represent the reflection of the tsunami from the east coast of Asia.

In fact, most tsunamis have oscillations that may continue for several days after the first waves strike. As late as 30 hours after the tsunami first hit Hilo, small tidal bores could still be observed advancing up the Wailoa River.

~

The characteristics of the 1957 Aleutian tsunami—speed, traveltime, wavelength, and distance from Hawaii—are very similar to those of the terrible 1946 Aleutian tsunami. In both cases the waves travelled over 2,440 miles in 4 hours and 55 minutes, a speed of just less than 500 miles per hour. In spite of these similarities, the effects of the two tsunamis were quite different. The submarine earthquake of 1957 was greater than that of 1946, yet the 1957 tsunami was much smaller and resulted in much less damage on the island of Hawaii. Kauai, on the other hand, was more severely damaged in 1957 than in 1946.

During the 1957 tsunami, only the northeast coasts of the islands recorded large wave heights; these waves averaged only about ten feet above normal, except in small, open bays, such as Pololu Valley where the water rose a prodigious 32 feet. In such areas, both underwater and abovewater topographies have an amplifying effect on the wave heights.

The complex interaction of the tsunami waves with the topography was also well illustrated in Hilo Bay. Eyewitnesses reported that the third or fourth waves were the largest on the west side of the bay, whereas the tide gauge located on the east side showed the third and fourth waves to have been the smallest of the first four waves. A seiche of the water in Hilo Bay is one explanation. As the water sloshed back and forth across the bay, it is possible that the waves on one side were amplified while those on the other side were diminished by the seiche. Observers at Cocoanut Island confirmed having seen a complex wave interference pattern produced near the middle of the bay.

The Performance of the Warning System

The Warning System once again proved its worth and not a single life was lost to the sea as a result of the 1957 tsunami. On the Big Island in Hilo, people were kept away from danger areas long before and after the large waves.

On the Island of Kauai there was much confusion on the part of the local authorities, in spite of the well-planned advisories from the Warning System. Yet, fortunately, even there no lives were lost.

What had been learned from the tsunamis of 1952 and 1957? One fact stands out above all others: each tsunami is unique. The Aleutian tsunamis of 1946 and 1957 showed that tsunamis coming from the same general place of origin may greatly differ in their comparative severity at any one site. This was well illustrated on the islands of Kauai and Hawaii, where the pattern of severity was reversed between the two tsunamis.

For Hawaii, the probable size of the tsunamis were predicted from wave heights reported for other locations and from the knowledge of past events. The specialists now began to see that many other factors must also be taken into account. The orientation of the coastline and the Hawaiian Ridge (the submarine foundation from which protrude the Hawaiian Islands) with respect to the direction of approach of a tsunami plays an important role in determining the average water heights along Hawaiian shores. Local topography, such as small, funnel-shaped bays, greatly amplifies the heights of the waves, whereas reefs provide the most effective screen.

In Hilo, the buffer zone provided by the coastal highway and park expanse proved to be an effective shield in protecting much of the town from the 1957 Aleutian tsunami. Yet compared with that of 1946, the Aleutian tsunami of 1957 was of only moderate proportions; would the barrier prove sufficient for a major tsunami or a tsunami originating in South America? Many questions remained to be answered.

The evaluation of potential danger of tsunamis to Hawaii is based on information from tide stations closer to the source of the tsunami. During the 1952 and 1957 tsunamis, the Warning System had admirably fulfilled the major function of reporting and evaluating the tide records and observations of water heights from the distant stations. Reports from the Aleutian and Alaskan stations and from Midway Island provided valuable data to the Hawaii warning center. In general, when a tsunami is generated in the North Pacific, stations in Alaska and Midway, as well as in Japan, the Soviet Union, Canada, and the continental United States, may all provide information on which to base a decision about the possible danger to Hawaii.

For tsunamis that come from areas south of Hawaii, the situation was not as clear cut. Prior to 1960, in the case of an earthquake in Peru or Chile, there was much less information available and the information itself was less dependable. A great deal of uncertainty was involved in trying to decide whether or not a destructive tsunami was headed for Hawaii. In such cases the Warning System had always erred on the side of caution. However, this situation is not always appreciated by the public. In 1958, a Mexican earthquake resulted in a tsunami alert. No tsunami was generated and the public considered the warning a false alarm. A well-located tide station could have immediately cancelled the alert, but none existed. In order to avoid unnecessary alerts, the Warning Center stressed the immediate need to establish more tide-observation stations between vulnerable areas and the probable sources of tsunamis. At the time, however, resources were just not available.

In 1960 the South Pacific was still without sufficient outlying tide stations. In the event that a tsunami originated along the coast of South America, confirmed advance warning would not be available. The situation was prophetically stated in a 1959 scientific paper on tsunamis (Fraser et al. 1959): "Hilo will be the outstation, and the only present recourse is a general warning without corroborative tide-gauge data." How would the population react to another tsunami alert? To another "false alarm"?

CHAPTER FIVE

Disaster by Night: The 1960 Tsunami

SUNDAY, May 22, 1960, was a day of terror for the South American country of Chile. Located between the high, rugged Andes Mountains and the great ocean depths of the Peru-Chile Trench, the city of Concepción had been rocked by sizeable earth tremors throughout the day and night, including a major earthquake measuring 7.5 at 1003 GMT on May 21. Then at 1910 GMT on May 22 another major earthquake with a magnitude of more than 7.5 shook the town. Only seconds later, as walls were still crumbling from the tremor, an even larger quake began at the same epicenter. For the next four minutes the quake intensified as movement was extended along the fault in the earth's crust. For an instant the shaking seemed to slacken somewhat, but then it increased with unbelievable force. Finally, after nearly seven minutes, it stopped. At its maximum, around 1915 GMT, it measured a colossal 8.5 on the Richter scale—more than 30 times the energy of the earlier 7.5 quakes.

From the time that the first earthquakes from Chile were registered at the Honolulu Observatory, the scientists at the Warning Center remained in a tense state of alert. The first major earthquake (Richter 7.5) at 1003 GMT on May 21 generated a tsunami that produced a small but noticeable wave in Hilo Bay. After the earthquake measuring 8.5 was recorded, authorities predicted that a tsunami had been generated—perhaps even a large, destructive tsunami.

If a tsunami were on its way, it would take nearly 15 hours for the waves to travel the 6,600 miles between Concepción and Hilo; the tsunami would arrive, therefore, around midnight Hawaii time.

During the early afternoon in Hawaii, news reports began to drift in from Chile. These reports told of destructive waves and damage along the coast of South America, confirming that at least a local, if not a Pacific-wide, tsunami had been generated. Geologists at the Hawaiian Volcano Observatory on the island of Hawaii waited anxiously, listening to the radio for further news.

Meanwhile, the scientists at the Tsunami Warning Center knew that a decision about a Tsunami Warning had to be made—a decision without the backing of reports from stations along the path of the waves between South America and Hawaii. Preferring to err on the side of caution, the Warning Center issued a Tsunami Warning at 6:47 P.M. The coastal sirens in the Hilo area began to sound at 8:30 P.M.

But just after 9 P.M., radio stations carried reports from Tahiti stating that the tsunami had reached Papeete and that the waves were an unspectacular 3 and a half feet high! Was this tsunami warning to be seen as another false alarm? How would the population react to the warning?

In Hilo most people had heard the news that a tsunami was supposed to arrive about midnight. But many people didn't really understand the warning. In fact, just a few months before, the actual system of warning sirens had been changed. Under the old system there were three separate siren alarms: the first siren indicated a Tsunami Warning was in effect; the second meant that it was time to evacuate; and the third was set to go off just prior to the arrival of the first waves. Under the new system, there was only one siren; it meant evacuate immediately!

After hearing the first siren, many people began to pack up their belongings in preparation for evacuation. Then they waited for the second siren before leaving their homes. There was to be no second siren that night!

The reaction of other Hilo residents, however, varied greatly, even among those who had firsthand experience of the wave in 1946. Some took it very seriously indeed. In Keaukaha, most of those who had seen the destruction of 1946 evacuated immediately. Chick Auld recalls how the police went from home to home, making sure that everyone was aware of the threat. Chick and his family left—so too did the Cooks and the Talletts. Paul Tallett wondered for a while if it were really necessary because there had been false alarms since 1948, but in the end he decided to be cautious.

Similarly, the staff at the Hilo Yacht Club left their workplace, wondering whether once again the building would be washed away. This time they took all the important documents with them.

All through Hilo, businessmen were moving documents and money from their vulnerable offices to safer ground. Although Bobby Fujimoto had relocated his Planing Mill *mauka* of Kamehameha Avenue after 1946, he decided it would be wise to remove his books and accounts. Many others took the same prudent action, but not everyone. Some were undecided. It was not that time had dimmed their memories of the shocking experience of 1946—they remembered with a fixed and awful clarity. But they could not believe that it was going to happen all over again, that this warning might need to be heeded when many in the past had been false.

Someone else with vivid memories of 1946 was "Baby Dan" Nathaniel. In 1960 he was working as a tour guide and had a party of tourists at the Hawaii Volcanoes National Park when the tsunami warning was given. He took his group back to their hotel—the old, wooden Naniloa—and told them to pack their bags. At 8 P.M. they were taken to the Hilo Hotel on Kinoole Street.

Just before 10 P.M., a group of geologists from the Hawaiian Volcanoes Observatory drove down to Hilo from the National Park. In spite of the darkness, they felt that they might be able to make useful observations of the tsunami. After arriving in Hilo and clearing their plans with the local police, they drove through the now-deserted streets to the Wailuku River. There they set up their observation post on the bridge overlooking Hilo Bay—the same bridge where just over 14 years before Jim and Bob Herkes had witnessed the awesome waves of the 1946 tsunami. The geologists kept busy measuring the heights above sea level of various reference points on the bridge pier opposite them. They intended to document the water level of each wave of the tsunami. They also planned their own evacuation route, a short sprint along the highway to safe, high ground. And when all was noted and ready, they waited.

~

The relatively small tsunamis of 1952 and 1957 had given some residents a false sense of security. The curious, the foolhardy, and the misinformed actually went down to the bay to wait for the waves to come in. They stood around the old Suisan Fish Market filled with the excitement and sense of adventure instilled by a late-night outing (Figure 5.1). All of Hilo waited for midnight and the new day, May 23, 1960.

~

Just after midnight the geologists observed that the water under the bridge had begun to rise. Within five minutes it reached a crest 4 feet above normal. Then the water slowly fell, and by 12:30 A.M. a trough 3 feet below normal was recorded. The first wave from the

Figure 5.1 Curious Hilo residents await the first waves of the 1960 Chilean tsunami.

Chilean tsunami had arrived. A few minutes later, word reached the geologists that the first wave had flooded the sidewalk near the bridge across the Wailoa River at the south end of the bay.

Reports from a Honolulu radio station stated that no waves had yet arrived at Hilo and that the estimated arrival time had been set back half an hour. Yet with their own eyes the geologists had seen the first wave pass. The radio report could only mean that communication between the Warning System and the news media had broken down. Misinformation had now increased the danger to the public.

At 12:46 A.M. the second crest washed under the bridge at a level 9 feet above normal. The wave topped the seawall of downtown Hilo and flooded the heart of the business district along Kamehameha Avenue. This wave was as large as the largest wave of the 1957 Aleutian tsunami.

Now the water began to withdraw again from the bay. At 1 A.M. the level measured nearly 7 feet below normal. At this point a strange calm prevailed, which the geologists described (Eaton et al. 1961):

> At first there was only the sound, a dull rumble like a distant train, that came from the darkness far out toward the mouth of the bay. By 1:02 A.M. all could hear the loudening roar as it came closer through the night. As our eyes searched for the source of the ominous noise, a pale wall of tumbling water, the broken crest of the third wave, was caught in the dim

light thrown across the water by the lights of Hilo. It advanced southward nearly parallel to the coast north of Hilo and seemed to grow in height as it moved steadily toward the bayshore heart of the city.

At 1:04 A.M. the 20-foot-high nearly vertical front of the in-rushing bore churned past our lookout, and we ran a few hundred feet toward safer ground. Turning around, we saw a flood of water pouring up the estuary. The top of the incoming current caught in the steel-grid roadway of the south half of the bridge and sent a spray of water high into the air. Seconds later, brilliant blue-white electrical flashes from the north end of Kamehameha Avenue a few hundred yards south of where we waited signalled that the wave had crossed the sea wall and buffer zone and was washing into the town with crushing force. Flashes from electrical short circuits marked the impact of the wave as it moved swiftly southeastward along Kamehameha Avenue. Dull grating sounds from buildings ground together by the waves and sharp reports from snapped-off power poles emerged from the flooded city now left in darkness behind the destroying wave front. At 1:05 A.M. the wave reached the power plant at the south end of the bay, and after a brief greenish electrical arc that lit up the sky above the plant, Hilo and most of the Island of Hawaii was plunged into darkness. (Figure 5.2)

Men like Oliver Todd who loved the ocean—swam and fished in it—could not imagine living away from it. Even though his home had been destroyed in 1946, he and his family had returned to live on the ocean front. Their new house had been built on land leased from the state (on the site of the present-day Hilo Hawaiian Hotel). On May 22, their son Oliver Jr. was home on leave from the navy. He and his father loaded some of their most valuable possessions in a truck, but decided to stay at the house and await events. Josephine Todd went to Villafranca (the area around Hualalai Street) in the evening. Like most people in Hilo, they listened to the radio. However, it was not from the radio that the Todds learned that the wave was an actuality rather than a possibility. The warning note was sounded by empty oil drums banging together underneath the house. When they heard the sound of the oil drums, they knew the water had risen behind them. In their heavy-laden truck they drove in the direction of Waiakea school, but were soon confronted by a flood of water rushing toward them. Without hesitation they took the only action available to them, and swung the sturdy vehicle straight through the bushes into the houselots subdivision.

Figure 5.2 The Waiakea Town clock stopped at 1:05 A.M. when the biggest wave struck. The clock now stands as a monument to the 1960 tsunami.

Not so lucky was Takeo Hamamoto. His home was in Liliuokalani Gardens (Figure 5.3), which he tended for the county. He had heard the warning and sent his son and daughter to safety with his brother. He stayed at the house with his wife and work crew, with the idea that they would save some of their possessions if the need should arise. Near midnight—the estimated wave arrival time—water began to flow into the park. Takeo decided that they should move his family's new washing machine to higher ground; however, just as they had done so they saw the white water of a wave breaking over Cocoanut Island. With all haste they piled into the car, but found the road flooded. How could they save themselves now? With the water surging behind them, they ran into the Waiakea schoolyard, where the crew climbed into the *hala* trees and Takeo and his wife clung to the large monkeypod tree. Building debris washed all around them—the water was up to Takeo's chest. As the crest of the wave passed, he and his wife climbed higher into the monkeypod tree. Hearts hammering with the exertion and excitement, they had another fright as a bright blue flash and loud explosion signalled to all of Hilo that the wave had reached the electric power plant.

The explosion of the power plant was the first indication for Tom Okuyama that the big waves had arrived. Working in his office at his Suresave Supermarket (then on Kamehameha Avenue), he had been listening to the radio but heard no news of the first waves. The water inside the supermarket was knee-deep, but he was able to make his way out through the back of the building.

Nearer to the Wailoa River, the force of the waves was greater. Evelyn Miyashiro had not urged her husband to move away from the area, despite her experience in 1946 when a tsunami wave had washed over her on the bridge. The restaurant they owned had done well; indeed, just 23 days before this night they had opened new premises, this time on their own property. It was not surprising, therefore, that they wanted to stay near both their business and their new home next door, in which they had lived for just one year. After the warning had been issued, Evelyn and Richard decided to stay with their property, since the flooding there had not been too severe in 1946. Together they watched people tying up their boats at the riverside. The tension present in each one was released by the wild cries of those who had been staring out to sea: "A big one is coming!" The

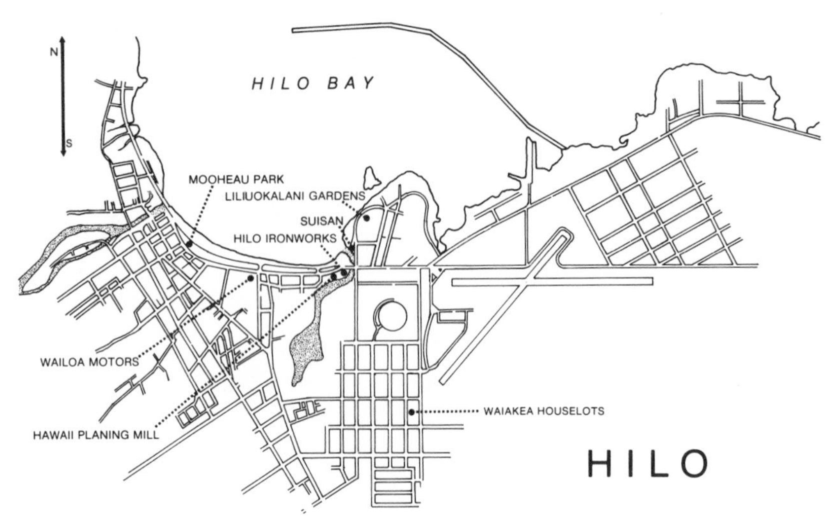

Figure 5.3 Map of the Hilo area showing selected locations mentioned in chapter five.

Miyashiros thought of trying to drive away, but could not find their 14-year-old daughter. They ran to their home where they found her praying upstairs. Then the whole family—mother, father, and three daughters—joined hands and prayed. As they prayed they felt the house being lifted from its foundations and carried along on the water. The lights went out and they had no idea where they were being carried. Evelyn was sure they were being taken out to sea, and that they were about to die. For what seemed a very long time they stayed in the room holding hands, until they saw the light of a flashlight and heard a voice calling. The house had become wedged on a piece of higher ground. The Miyashiros felt themselves blessed in having been saved, even though their new restaurant was destroyed. For the second time the sea had attacked their business, this time more fiercely than before.

Tadayoshi and Sushisako Okamoto's store on Mamo Street was also washed away by the big waves. They had repaired the store after the 1946 tsunami, but moved their residence from the apartments above the store to a house on Piopio Street. The family was asleep in their home when the waves began. They had heard the warning earlier, but thought it was just another false alarm. Mrs. Okamoto had been at the beach that afternoon with her children, and observed that the sea was calm, with no hint of the wave travelling at high speed across the Pacific toward them. The first they knew of its arrival was a "roaring like an express train." Their reaction was to spring from bed and turn on the lights, but as soon they had done so the lights were extinguished by the wave's assault on the power plant. Realizing that it was too late to run, Mr. Okamoto put a mattress over his wife and three children to protect them from debris should the walls collapse. But like the Miyashiros, they were among the lucky ones. When the biggest wave had passed, they waded through the flooded ground floor of the house to their backyard. With superhuman strength Mr. Okamoto was able to wrench open the heavy gate, and in their night clothes and bare feet the family fled the area and ran *mauka* as fast as they could.

Mark Olds lived near the Todds, on the site now occupied by the Hilo Bay Hotel. He was one of the many residents who evacuated the danger zone, only to return before the danger had passed. Although he heard a tsunami warning on the radio around 2 P.M., he did not leave his home until 4:30 P.M. when another warning was broadcast. In his office in the downtown district he continued to listen to his radio. As he remembers, he did not hear any statement of the expected time of arrival of the first waves. He did hear that the tsunami had reached Tahiti, where the height of the biggest waves was no more than 3 and a half feet. On hearing this information Mark decided to return to his house, where he spent the evening watching television. No warnings were broadcast by the television station. Around midnight he was ready to go to bed, but on impulse he thought he would take a look at the ocean before retiring. Opening his back door, he switched on a yard light which revealed that the yard was completely flooded. Mark decided it was time to leave, and moved quickly to the front of his house which was at a higher elevation. As he reached the *lanai,* the explosion from the electric plant reinforced the urgency of his situation. Through the blackness Mark felt his way to the car. In those days he kept his keys in the car, behind the visor. But in the dark his hand hit the visor, and the keys fell to the floor. The next minutes, which seemed like hours, were spent scrambling with shaking fingers over the car floor. Above the pounding of his heart, Mark could hear the splintering of wood and rushing of water. At last he found the keys and started his car. Just as he reached the street, however, another driver came along and parked across his driveway. The newcomer left his car, ran across the street to the park, and began to climb a utility pole. Mark ran to the foot of the pole and shouted up to the climber to move his car. "Forget the car!" commanded the voice from above, "the wave is coming *now*. Get up a pole!" Obediently, Mark began to climb, only to be stopped in his ascent by frenzied shrieks. "Not this one, not this one, go to the other pole!" Too bemused to argue, Mark jumped to the ground, ran to the next pole, and scampered up it just ahead of the rising water. He climbed about three-quarters of the way to the top—the water reached his feet. When it had subsided, Mark and his new acquaintance left their perches and drove off in their respective cars. This proved to be less than a good idea for Mark. As he drove into the Waiakea district the water rose again, causing his car to stall. Hearing calls for help, he left his car and went across to a house to assist a woman who was trying to escape from her upstairs window. The woman landed on top of him when she jumped from the window—immersing them both in the muddy flow. They rose from the flood, however, and were able to make their way on foot to higher ground.

In spite of their terrifying experiences, people like Mark Olds and the Okamotos remained where they knew there was land underneath them. But there were others who would feel the full power of the ocean, as Herbert Nishimoto had at Laupahoehoe in 1946.

The 1946 waves did little damage on the Puna side of the Wailoa River. Houses along the river had been flooded on the ground floor, but none had been destroyed and no lives were lost. In fact, on that occasion many residents from the neighborhood had run to the base of the big mango tree near Fusayo Ito's home. Because of this experience, Mrs. Ito and many others decided to stay in their homes. Mrs. Ito's recollection of 1946 was one of excitement rather than terror. She remembered hearing the approach of the waves and thinking it was the noisy baseball team from Honolulu, just disembarked from the boat! And so, despite her daughter's pleas that she should leave, on this occasion she opted to stay. Mrs. Ito would have liked to have gone down to the river bank like most of her neighbors, but she didn't want to venture out in the dark. Instead, she watched from her door until shortly after midnight, when people began to walk by saying that the time of danger had passed and nothing had happened. For a few moments she felt a great relief of tension—then her world was shattered. Her heart leapt as she heard the alarming sound of an explosion "like a bomb," and she was enveloped in darkness. In the next instant the wave entered her open door, seized her and spun her around and around, and churned everything in her home. She was hit on the head, fell through a hole, struggled to lift herself, and lost consciousness.

The next she knew was the sensation of being among bushes. How could bushes be in her house? Maybe her house had been moved to the river's edge. Eyes tightly closed in fear, she became aware of the sound of water. She moved one leg and tried to find a foothold—but met nothing solid, only water. For a while she floated on her back, then slowly opened her eyes. Above her in the great black expanse shone the stars. "Then that's the first time I cried—cry, cry, cry, because I was so, so scared—nobody around." Then she was deluged by another wave: in the thick darkness she swallowed more water, smelt gasoline, and was whirled around and propelled toward the bay. Borne up by a piece of debris, she heard whistling nearby and a man's voice calling "Can you swim?" "No" she replied. "Then hang on!" Grasping the piece of debris, she was dragged by the ebbing water past the Hilo Ironworks and between the tops of two large pine trees. Moving very fast, Mrs. Ito was washed down by Wainaku mill, saw its light, then found herself in the ocean.

Her eyes had now adjusted to the darkness, and she could see the vast amounts of debris that surrounded her. In her confused state she thought the surface was so thickly covered with wreckage that she would be able to walk back to the shore. She put one leg over the edge of her makeshift raft, and realized that the water was very deep. She could see the lights of vehicles moving along the shore, but knew she could not reach them. All alone on the ocean, she heard no sound but the roaring of breakers. About this time, she became aware that she was being supported in the ocean by a window screen from her house. Only a tiny woman like herself—a mere 4 feet, 11 inches tall —could have been kept afloat by such a flimsy structure. All night she was tossed on the turbulent water, lifted on the crests of the waves, then plunged precipitously into the troughs. By this time no land was visible, but she saw some lights far away. Then there was only "sky and ocean, sky and ocean." Mrs. Ito cried for a while, then made peace with her God. The sky was beautiful. She accepted that death would come eventually, by sharks or by drowning, but she felt no concern. She had no control over whatever might happen.

By 2:15 A.M. the height of the waves reaching the bayfront had diminished, and the geologists from the volcanoes observatory felt it would be safe to enter downtown Hilo to assess the damage. They were alarmed at what they found (Eaton 1961):

> Thick slimy mud covered the streets, and fish abandoned by the water that carried them over the sea wall were strewn about. Hilo's sewage, dumped inside the harbor entrance, had been stirred up by the first two waves and hurled into the face of the city by the third, filling the air with a distressing stench. At the north end of Kamehameha Avenue damage was slight, consisting only of broken windows and muddied floors. Stores in the block north of Haili Street had been breached by the waves, which gathered up their contents and dumped them in confusion on the street. Broken power poles, tangled wires, yardage goods festooned through wires and muck, children's toys and gasping fish clogged the street, making our progress southward treacherous and slow.

As the geologists continued to work their way through the devastated town, they saw four people emerge from a second-story window in one of the few buildings left standing. The stairs had been carried away as the waves had gutted the ground floor. The geologists helped the survivors down to the street. "Only then, as we picked our way northward along Kamehameha with our unexpected charges, did the horrible reality sink home: Hilo's streets had been evacuated, but its buildings had not!"

~

Rescue operations continued throughout the night and into the early-morning hours. A major concern was the number of people missing, especially given the possibility that many had been washed out to sea by the giant waves. Indeed, there were such victims: as Monday, May 23 dawned in Hilo, Mrs. Ito was still alive, still floating offshore.

The morning tide carried her back toward Hilo. As the light spread across the water, she saw that most of the debris had dispersed during the night and she was alone on her window screen, except for a large wooden platform floating behind her. Eventually she saw something white, and wondered if it could be a "ghost ship," a figment of her imagination. In fact, it was the Coast Guard's 95-foot patrol boat under the command of Chief Boatswain Fredrick R. Nickerson. Mrs.

Ito had been spotted about 800 yards from the boat. As it approached her, two sailors jumped into the water and swam to her. When Fusayo Ito heard the splashes and saw her rescuers, all the peace and resignation she had felt during the night left her. First she was rigid, then she surrendered herself to their ministrations; she was lifted aboard, given first aid, and wrapped in a blanket. After her long ordeal her only physical injuries were a cut finger and bumped knee. But the effects of the shock were great, and for many weeks the sound of water would set her shaking.

Although the 1960 Chilean tsunami caused only moderate damage on most of the Hawaiian Islands, in Hilo it had been catastrophic. The first wave had reached Hilo at 12:07 A.M. local time, and as in 1946, 1952, and 1957, the first indication was a rise in the water level. The tsunami travelled the 6,600 miles from Chile in 14 hours and 56 minutes—an average speed of 442 miles per hour. The arrival time predicted by the Warning Center had been 20 minutes earlier than the actual arrival time in Hilo, an error margin of only 2 percent.

The two tide gauges in Hilo Harbor were put out of action by the waves, so the best record of the movement of the water in Hilo Bay during the tsunami was provided by the observations of the geologists from the volcanoes observatory. They recorded a period between waves of just over half an hour for the first two waves, but the third wave was very different. Ten minutes before the third crest should have washed into Hilo, a giant vertical wall of water—a bore—advanced on the town.

The popular myth that a great withdrawal of water from the shore precedes a giant tsunami wave may be more fact than fiction. A tsunami wave may be transformed into a bore when it advances at sufficiently high speed through very shallow water. The formation of the bore in Hilo Bay may have been initiated by the large withdrawal of water during the trough of the second wave. Descriptions from observers near the Wailuku River and reports from boatmen near the harbor entrance indicate that before the bore arrived, the water in the harbor was at a level about 7 feet below normal; the bore formed initially at the harbor entrance, near the end of the breakwater, where the depth of the water becomes shallower.

The necessary condition for the formation of a bore is that the advance of water back into the bay following the previous withdrawal must be faster than the "critical" velocity* for shallow-water wave motion. Estimating the depth of water remaining in the bay to be around 5 meters (approximately 16 feet) would give a critical wave speed of about 7 meters per second, or just over 16 miles per hour.

As the crest of the arriving wave moves into shallow water, it slows down. Some of the energy of the advancing tsunami is lost due to increased friction with the shallower seafloor and some of the wave's energy is converted into increased wave height. The wave's speed is now controlled by the depth of water through which it is moving. Since the depth is always greater (by the height of the wave) at the crest than at the trough, there is a tendency for the crest to "catch up" to the trough, and the wave front becomes steeper. Usually this action causes the wave to break. However, when the factors of bottom slope, speed of the out-flowing current, and height and speed of the incoming wave all combine appropriately, a nearly vertical wall of water, moving rapidly toward shore, can be produced.

If the speed of the incoming wave crest (in reference to the third wave of the 1960 tsunami in Hilo Bay) was greater than this critical value of 16 miles per hour (and it almost certainly was), the advancing wave would break free of the water ahead of it and below it, and a bore would be produced. When this occurs, the speed of the wave actually increases as it advances toward shore (the exact opposite of the behavior of ordinary waves in shallow water, which travel more slowly as the depth decreases). Evidence suggests that the height and turbulence of the bore reached a maximum in the bay between the harbor entrance and the Hilo bayfront, like enormous breaking surf. At this point, it was estimated to have been travelling at more than 30 miles per hour. The wave height along the shore steadily increased southeastward along the bayfront, reaching a towering 35 feet near the Wailoa Motors building, located at that time on Kamehameha Avenue. The final flood of water into Hilo was described by scientists as being "analogous to the sheet of water that races up a beach beyond the spent breaker that propelled it" (Eaton et al. 1961).

As the wave surged into Hilo, it wrenched 22-ton boulders from the 10-foot-high bayfront seawall and carried them as far as 600 feet inland across Mooheau Park, without leaving a noticeable mark on the lawn. The water struck with such force that 2-inch pipes support-

*This critical velocity for shallow-water wave motion is given by the formula: $C = \sqrt{gh}$ (where C is the wave velocity, g is the coefficient of gravity, and h is the depth of water).

ing parking meters along the waterfront were bent over parallel to the ground (Figure 5.4). Electric cables and transformers were torn from utility poles. The reinforced concrete office of the Hilo Iron Works withstood the force of the wave, but its second-story skylights were blasted out by the increase in air pressure as the wave struck the building—it had become a manmade "blowhole." At a nearby showroom, an 11-ton tractor and the building housing it were removed by the wave.

Figure 5.4 Hilo bayfront parking meters, bent parallel to the ground by the tremendous force of the waves of the 1960 tsunami.

Bobby Fujimoto's Hawaii Planing Mill lay directly inland of the maximum 35-foot wave height measured along Kamehameha Avenue. The large, modern hardware store made of prefabricated steel was completely swept away, right down to its tiled cement floor. The only remnants were half-inch steel reinforcing rods sticking up from the foundation. Pieces of the building were later found 1,500 feet away beyond the Wailoa River.

In this zone of total destruction, entire city blocks were swept clean. Buildings were wrenched from their foundations and deposited as piles of debris hundreds of feet away. Some light frame structures were floated off their foundations and deposited elsewhere, sometimes in clusters, without major damage except by collision with other structures. Others were pulled into Hilo Bay, some drifting offshore and breaking apart—littering Hawaii's shores for weeks afterward.

Scores of automobiles and trucks were totally wrecked (Figure 5.5). In some cases cars were stacked three-deep by the waves. Thirty-foot lengths of concrete curbing from the Bayfront Highway were carried 350 feet inland. And once again, the strong footbridge to Cocoanut Island, replaced after the 1946, 1952, and 1957 tsunamis, was swept away. All of this damage was done by the third wave.

Even to this day, scientists are not completely sure exactly how the bore formed and arrived ten minutes before the third crest. One suggestion was that a harbor seiche (the natural surge of water in and out of the bay) had acted in concert with the advancing crest of the third wave. But the arrival of the bore ten minutes before the third crest argues against this hypothesis. It was more likely that the harbor seiche added to the withdrawal of water during the second trough, which then lowered the water level and set the stage for the bore to form. In any event, after the third wave the period between waves was 15 minutes rather than 34 minutes, the same periodicity as occurred in the 1952 and 1957 tsunamis—probably representing the natural period of oscillation of Hilo Bay.

On the other Hawaiian Islands the tsunami reached moderate heights and caused moderate but widespread property damage. With the exception of Hilo Bay, the wave heights on the island of Hawaii itself rarely exceeded those of the 1957 Aleutian tsunami, and averaged only 9 feet. There were some local variations in wave heights, mostly in V-shaped inlets or shallow bays, with 17-foot waves recorded at Honuapo and Kaalualu, and 12-foot waves at Honomu.

Figure 5.5 Mail truck smashed into a house, part of the damage caused by the tsunami of 1960.

Along the west coast of the island a wave height of 12 feet was reported in the small bay at Keauhou, and a height of 10 feet reported at Kahaluu. Only at Napoopoo were the waves of the 1960 tsunami larger than those of 1946 or 1957: waves 16 feet high washed over the settlement, destroying six houses and moving a number of others off their foundations. Eyewitnesses told of waves coming from the north and washing southward over the low-lying village. The movement of buildings and debris also indicated that the waves came from the north, not from the south—the direction of Chile. How can this be the case? Napoopoo is situated on the south shore of Kealakekua Bay and lies opposite the steep cliffs of the Kealakekua fault zone, which form the north shore. The destructive waves were, perhaps, the result of the main tsunami waves (coming from the south) bending into the bay and being reinforced by the reflected tsunami waves bounced off the cliffs. This combination produced the 16-foot waves which washed over the village.

As in 1946, the city of Hilo suffered the most extensive damage in all of the Hawaiian Islands. The business district along Kamehameha Avenue and the adjoining low-lying residential areas of Waiakea and

Shinmachi were literally wiped off the map. Damage to property included 229 dwellings and 308 businesses and public buildings. The floodwaters inundated approximately 580 acres between the Wailuku River and the shoreward end of the breakwater. Between the Wailoa and Wailuku rivers the water washed inland as far as the 20-foot contour above sea level. Property damage was estimated as high as $50 million.

But the greatest cost could not be measured in dollars. In Hilo, 61 people were crushed or drowned by the tsunami waves, and another 43 people were injured, requiring hospitalization and medical care. Fortunately, no lives were lost elsewhere in the Hawaiian Islands.

Once again, Hilo had been brought to its knees. Until the water mains were repaired, residents were urged to boil their water. The pumps of the local sewage system had been destroyed, leaving many areas with no sewer service. The destruction of the electric power plant left the electric company unable to meet the power needs of the island. The County Health Department advised that damaged food stocks be inspected before use. Milk had to be brought in from Honolulu because the local pasteurization facilities had been destroyed. "Fog" machines were brought in to spray insecticide in potential mosquito breeding areas. Public shelters had to be used to house 215 families until other arrangements could be made.

~

Like many other Hilo residents, Mrs. Ito lost her home in the deluge of the Chilean tsunami. She had no recompense and asked for none, thankful to be alive. Her life savings, in the form of savings bonds, had been swept away with her home. But this part of the story was to have a surprise ending. All her important papers, including the bonds, were kept in a waterproof bag. When clearing work was being done on the side of the river opposite the site where her house had stood, the bulldozer driver was stopped by an obstruction. When he went to clear it he found a waterproof package, which he then took to the police station. What a wonderful surprise when the police called Mrs. Ito! "I cried and cried—if I was dead, I don't need those things, but if I am alive, I need them." Earlier, she had tried to claim the value of the bonds from the bank, but had not known the serial numbers. It took her days to soak the mud from the bonds, but in the end the serial numbers were visible, and the bonds were honored by the bank. Buddha was looking after her. The sum of her experiences makes her feel that she was saved in order to help others.

CHAPTER SIX

The 1960 Tsunami: What Went Wrong?

THE 1960 Chilean tsunami, like that from the Aleutians in 1946, was a major catastrophe for Hilo. In the wake of the destruction, the authorities began to assess the cost of the disaster as well as attempt to determine why there had been 61 deaths. It was time to figure out what had gone wrong with the system and what could be done to prevent another tragedy in the future. It was also time to rebuild.

The Kaiko'o Project

Following the 1946 tsunami, the strip of land between Kamehameha Avenue and the bayfront was converted into a recreation and parking area that was to serve as a buffer zone for future tsunamis. Many of the businesses displaced by this buffer zone were rebuilt in other low areas. Some of the residential communities that had been severely damaged in 1946 were rebuilt with crowded, flimsy houses. The impact of the Chilean tsunami waves on these residential and business areas, particularly at Waiakea and Shinmachi, was the cause of much of the damage and loss of life in 1960.

The events of the 1960 tsunami forced the realization that Hilo would always be at the mercy of destructive tsunamis. Just eight days after the tsunami, the Hawaii Redevelopment Agency was established. This agency was responsible for the extension of the oceanside buffer zone and the construction of a landfill plateau, which raised the inland border of the greenbelt 26 feet above sea level. The word *kaiko'o,* Hawaiian for "rough seas," was chosen as the name for the project, and it soon became the new commercial center of Hilo. Federal and state funds for public housing and urban renewal were provided, and the Small Business Administration made loans available to help businesses get started again. Because the local population was justifiably skeptical about rebuilding—even behind the buffer zone—buildings for the state and county were the first to be constructed. They now stand on the bluff overlooking the buffer zone, just seaward of the busy Kaiko'o shopping mall.

The Impact of the Warning System

The Tsunami Warning System was, perhaps, the greatest success yet also the greatest failure of the 1960 catastrophe. The system had warned of the approach of the waves as well as accurately predicted their arrival times, but it was apparent that public education as to the nature and seriousness of the tsunami had been totally ineffective.

During the 1952 and 1957 tsunamis, large numbers of people failed to leave danger areas when told to do so. Instead, sightseers converged on the coast. Thanks to the small size of the waves, there was no loss of life. But this display of curiosity should have served as a warning to the authorities that the next large tsunami would kill many people. After the 1960 tsunami, newspaper headlines indicated that Civil Defense officials were shocked by the reactions of residents. Yet people reacted no differently than they had in 1952 and 1957.

Many people returned to danger areas after the first small waves had passed, only to be overwhelmed by the giant bore. At Kuhio Beach on Oahu, people who should have known better ran out on the reef and picked up fish. Police set up roadblocks, but were unable to prevent sightseers from entering dangerous beach areas. At the Moana Hotel in Waikiki, guests refused to leave the Kamaaina Bar despite warnings from the management. The guests were lucky and survived, even though the bar was inundated by the waves. State Governor William Quinn pleaded with residents over the radio to remain on safe ground, but residents and visitors alike flocked to the beaches to witness the tsunami.

A study of the behavior of Hilo residents during the 1960 tsunami revealed that almost all of the 329 adults interviewed had received some kind of warning that a tsunami was forthcoming (Bonk et al. 1960). Yet only 32 percent evacuated the area after they received the warning. More than half simply waited at home for more urgent or specific instructions until it was too late and the waves struck.

Just after the tsunami struck, Governor Quinn declared a "state of emergency" throughout the state, which empowered the National Guard to maintain order. Later, it was debated whether or not the National Guard should have been used before the tsunami struck to enforce orders to keep the public out of danger areas. Oddly enough, it was thought that this might infringe on a citizen's "constitutional right" to endanger himself by remaining in a threatened area.

The most serious problem, however, is that the public had only vaguely understood how the warning was to be given and how they were expected to respond to it. After 1960, the warning procedure was changed and the alarm system currently in use in Hawaii was established. The procedure for alarm warnings is as follows:

> Three hours before the arrival of the first wave, the Civil Defense sirens will sound the *attention/alert signal* (a three-minute steady siren tone, repeated as necessary). The *attention/alert signal* means "turn on your radio." Radio stations switch to State Emergency Broadcast System (EBS) status and regular announcements are made about procedures to follow, and current information on the tsunami is given. The *attention/alert signal* is sounded again two hours, one hour, and thirty minutes before the estimated arrival time of the first tsunami waves. Each of these four signals is accompanied by emergency announcements over the EBS.

Both the State EBS and the tsunami evacuation areas are described in the front pages of the telephone directory. The Civil Defense sirens are sounded as a test on the first workday of every month at 11:45 A.M. But if the population remains unaware of what the sirens mean or fails to respond to the warning, the next destructive tsunami could well be another killer.

Tsunami Research Efforts

The 1960 tsunami gave scientists another firsthand opportunity to study how tsunami waves behave in the Hawaiian Islands. Immediately following the tsunami, researchers set about collecting data. The precise directions of wave advance were determined by measuring what had been wrought by the tsunami's force: the "compass bearings" of bent parking meters and traffic signs; the direction of gouge marks left by buildings and other heavy debris; and the orientation of lines between the original location of buildings and where they were deposited by the waves.

The heights reached by the waves were calculated by noting the level reached by the water, as indicated by various kinds of evidence: salt-killed grass and other vegetation along the shore; persistent strand lines of fresh cane trash; abraded bark and broken tree branches; debris left hanging on fences, buildings, and trees; and water stains and abrasion on buildings. The distance of each of these features above sea level was carefully measured and then later reduced to the distance above the mean lower low water datum level.

The results of the research show a striking contrast between the 1960 Chilean tsunami and the 1946 Aleutian tsunami. The waves from the 1946 Aleutian tsunami reached a maximum height of 55 feet and averaged a height of 30 feet along the northeast coast of Hawaii. The waves from the Aleutian tsunami travelled directly to the northeast coast—the heights there averaged more than twice that of the island as a whole. A similar increase in wave height might have been expected along the southeast coast in 1960; this is the side of the island that most directly faces Chile. Yet in 1960, the waves were no larger on the southeast coast than along the more-protected west and northeast coasts. Specialists believe that these variations in wave pattern might be caused by the difference in cross section of the Hawaiian Ridge as encountered by waves approaching from different directions. The ridge presents a barrier of almost continental dimensions to Aleutian tsunamis, which approach it broadside, but only a small barrier to Chilean tsunamis, which approach it end-on. Another factor that might have led to larger waves along the northeast coast in 1946 is the relatively shallow, sloping feature called a shelf that extends outward underwater from the sea cliffs of the Hamakua Coast north of Hilo. This shelf feature is absent elsewhere on the island.

The 1946 and 1960 tsunamis also show a marked difference in their pattern of flooding in the Hilo area. In the central part of downtown Hilo (northeast of Kumu Street), and east of the Waiakea Peninsula, the 1946 waves flooded inland farther than the 1960 waves (Figure 6.1). The 1946 waves were relatively high (17 to 25 feet) throughout the entire bayfront area from the Wailuku Bridge to the Waiakea Peninsula, but were lower in the lee of the peninsula. In 1960, the waves were large even in the lee of the peninsula. In fact, between downtown Hilo and Reeds Bay, flooding was almost twice as extensive in 1960 as in 1946.

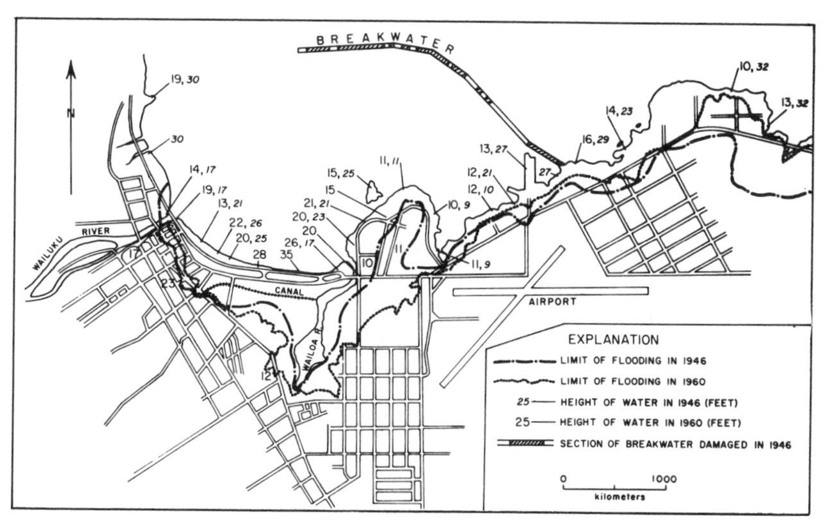

Figure 6.1 Inundation limits in Hilo of the 1946 tsunami compared to the 1960 tsunami.

The scientists concluded from their studies that although each tsunami is indeed unique, the location of a tsunami's source may be one of the most important factors in determining which area will be most affected by the waves, and how great the subsequent damage will be. Tsunamis from near the same geographic place of origin tend to produce remarkably similar relative patterns of wave height at any one location. For example, the tsunamis of 1946 and 1957 both originated in the Aleutian area, and although the maximum wave heights in Hilo from the two tsunamis were very different, the most serious damage caused by each tsunami occurred in the same parts of town. The 1960 Chilean tsunami, on the other hand, produced a very different pattern of inundation compared to either of the Aleutian tsunamis. In other words, on Hawaii's shores the patterns of wave heights and flooding produced by tsunamis of different geographic origin are strikingly different. Tsunamis from nearly the same origin, although perhaps differing in relative severity from place to place, tend to be more similar than previously thought.

The scientists conducting the investigations also learned more about how tsunami waves interact with coral reefs and small, steep islands. The radio announcements in 1960 of the small wave heights at Tahiti gave a false sense of security to many in Hawaii. Tahiti is surrounded by coral reefs. Reefs tend to break up the tsunami waves, dispersing and absorbing their energy. As one scientist put it, "Tsunamis can go past day after day and they'll hardly even know it."

Small, steep islands also tend to be minimally affected by tsunami waves. One of the Line Islands, located about a thousand miles south of Hawaii, is named Christmas Island. It is one of the very few tide stations lying between the South American coast and the Hawaiian Islands, and as such could have served as a valuable indicator of the severity of a tsunami heading toward Hawaii. The 1960 Chilean tsunami produced a 6-inch rise on the tidal record at Christmas Island. In 1960 this was interpreted to mean that there was a tsunami, but that it was small. As one of the scientists said, "We later found out that a 6-inch rise at Christmas Island is a big tsunami! These are the things that you're only going to learn after they happen. This is how we refine the art, so to speak. The more tsunamis we have, the better we'll get." Scientists at the Warning Center had learned painful lessons from the 1960 tsunami.

So too had the inhabitants of other Pacific islands. On Pitcairn Island, New Guinea, and New Zealand, and farther north in the Philippines and on Okinawa, some 300 lives were lost. In Japan, the Chilean tsunami waves had crashed ashore on the island of Honshū, killing over 100 people.

The great destruction caused by the 1960 Chilean tsunami prompted Japan and a large number of other countries and territories to become members of the Tsunami Warning System.

In Hawaii, the Hilo Bay area proved to be the most vulnerable area of the islands. The Chilean tsunami was bent, bounced, and funnelled into the bay with disastrous results. The 1952 Kamchatka tsunami radiated around the island and reached its maximum heights in Hilo Bay. Hilo Bay faces directly toward the Aleutians, the source of the 1946 and 1957 tsunamis. What would a truly large earthquake in Alaska do to Hilo?

The Earthquake at Prince William Sound

At 5:36 P.M. on Good Friday, March 27, 1964, one of the largest earthquakes ever recorded in North America struck Alaska. The quake was centered near the eastern shore of Unakwik Inlet in northern Prince William Sound, a sparsely settled region of high, rugged topography and numerous glaciers. When the magnitude of the

Figure 6.2 Damage to port facilities at Seward, Alaska.

tremors was later determined, they would measure an awesome 8.4 on the Richter scale, releasing twice as much energy as the famous 1906 earthquake that destroyed much of San Francisco.

The earthquake was produced by movement along a system of complex faults in the earth's crust that dip from the Aleutian Trench beneath the continent. The center of the movement occurred at a relatively shallow depth of about 14 miles. The quake was accompanied by vertical uplift or subsidence over an area of 170,000 to 200,000 square miles, which ranged from a drop of the land surface by 7.5 feet to a rise by as much as 38 feet and included the area of the islands and mainland of Prince William Sound.

The populated areas nearest to the epicenter include the small communities of Valdez, 40 miles to the east, and Cordova, 70 miles southeast, and the city of Anchorage, 80 miles to the west (Figures 6.2, 6.3).

Many Alaskans described feeling seasick from the motion of the earth as the seismic waves radiated from their source. In Anchorage, cars, trucks, and even aircraft were observed bouncing as though they were "on a trampoline." At Slana, Alaska, waves in the earth—esti-

Figure 6.3 Damage in Anchorage, Alaska, after the earthquake of March 27, 1964.

mated to be 18 inches high and 50 to 60 feet apart—were seen moving across the ground.

These powerful seismic waves set up seiches in rivers, harbors, channels, lakes, and even swimming pools as far away as Puerto Rico. In the New Orleans industrial canal in Louisiana, a 6-foot wave tore an 83-foot U.S. Coast Guard vessel from its moorings. In Texas, the Coast Guard reported that a surge came through the Sabine Pass on the Gulf of Mexico and pushed the tide 3 feet higher than normal. At the Tropicana Swimming Club in the Corpus Christi area, a 2-foot wave splashed over the edge of the pool, causing the loss of 25,000 gallons of water.

The seismic waves generated by the earthquake travelled the 2,400 miles from Alaska to Hawaii in just eight minutes—reaching the Honolulu Observatory at 5:44 P.M. (0344 GMT). The observatory staff were in their quarters for their evening meal, when all of a sudden the seismic alarm sounded. The scientists rushed back to the observatory and immediately noted the large earthquake trace on the seismographs. At 0413 GMT they sent out messages requesting seismic data from other observatories. Six minutes later, the first report came in with seismic readings from Manila. Within the next 20 minutes, reports had arrived from Hong Kong, Guam, Japan, California, and Arizona. Yet there were no reports from the observatories in Alaska at College and Sitka. Data from these observatories, as well as tide-station data from Kodiak, Sitka, and Unalaska, were routed via communications channels passing through the control tower at the Anchorage International Airport: what the scientists at the Honolulu Observatory did not know was that the control tower had been demolished by the earthquake—breaking the vital communications link.

By 0452, enough information had been reported by various seismic observatories to permit the Honolulu Observatory to locate the epicenter at 61° North, 147°30′ West, near Prince William Sound, Alaska. At 0502 the following advisory message was sent out to all agencies in the Warning System: "A severe earthquake has occurred . . . It is not known if a sea wave has been generated . . . If a wave has been generated its ETA [estimated time of arrival] for the Hawaiian Islands (Honolulu) is 0900Z [GMT] 28 March."

At 0530 the Warning Center issued another bulletin which stated: "Damage to communications to Alaska makes it impossible to contact tide observers. If a wave has been generated the ETAs are: [a list of some 33 sites around Alaska and the Pacific basin followed]."

In Alaska, tsunamis were already wreaking havoc. In fact, tsunamis from two different sources struck some areas. In many harbors near the epicenter, the seismic waves caused submarine landslides to occur. These landslides in turn generated highly destructive local tsunamis. Meanwhile, the main movement of the seafloor generated a major tsunami that would be felt all around the Pacific basin.

Locally destructive tsunami waves generated by landslides occurred at Valdez, Whittier, and Seward. At Port Valdez, a slide generated a wave that slammed into the waterfront within two or three minutes of the onset of the earthquake. The M.V. *Chena,* a 10,815 ton coastal freighter, was unloading cargo at the Valdez dock when the earthquake struck. The freighter's captain, Merrill D. Stewart, stated that a wave raised his ship some 30 feet above the pier which heeled the vessel over by 50 to 70 degrees into a "whirlpool" of mud and debris where the pier had been. According to the captain, "the ship was slammed on the bottom and rolled wildly from side to side like a rag doll" before it finally slithered off the mud bottom through the wreckage of the docks to reach the safety of deep water. The waterfront of Valdez was totally demolished—sand and silt were splashed as high as 220 feet above sea level by the waves. In the end, 31 people lost their lives to the slide and waves.

At Whittier, locally generated tsunami waves reached a height of 104 feet above sea level, destroying two sawmills, the railroad depot, and the Union Oil Company tank farm. On Homer Spit, the small boat harbor disappeared into a "funnel-shaped pool" within two minutes after the earthquake started. At Seward, waterfront oil storage tanks were ruptured by the earthquake and the petroleum immediately caught fire. A slide-generated wave reaching a height of 30 feet was followed half an hour later by another wave, probably the first of the major tsunami series. These waves—literally waves of fire—must have been a terrifying sight as they surged into Seward carrying the flaming oil. The railway docks, the electrical generation plant, and houses all caught fire and were destroyed. In the end, there were 12 fatalities and the entire economic foundation of Seward was destroyed.

Although the landslide-generated tsunamis were locally disastrous, the major tsunami waves ultimately caused the most death and destruction.

At the Warning Center in Hawaii, attempts were being made to obtain tide reports from stations near the epicenter in order to confirm the generation of a tsunami. But the combination of the violent earthquakes and the battering by tsunamis had left south-central Alaska without a single tide gauge in operable condition. The tide gauge at Seward would later be found on the deck of a tanker that had been moored at the dock when the earthquake struck. In fact, the only good record of the tsunami along the Gulf of Alaska was made by personnel of the U.S. Navy Fleet Weather Station at Kodiak. The city of Kodiak and the naval station were the only places in Alaska that received any advance warning of a possible tsunami, and they received word only ten minutes before the first waves—cresting 22-feet high—surged ashore. By timing and marking the crest heights of successive waves, and by estimating the ebb levels and times, navy personnel were able to construct a record of the tsunami waves. This information was passed on to the Warning Center with great difficulty. Because the earthquake had knocked out all radio circuits in the building, a telephone message had to be relayed to a remote navy radio station which then sent the message to Honolulu via San Francisco. When the first wave flooded the building all electric power was lost; however, within 15 minutes emergency power was supplied and navy personnel were able to send a follow-up message with more details on the first wave.

In the city of Kodiak, 21 people were killed by the tsunami but the toll could have been much worse. There were a few minutes of warning before the first wave and, fortunately, the wave was a gentle flood followed by a gradual ebb. Many were alerted by this wave and fled to safety. The second wave came crashing in as a 30-foot wall of water, washing 100-ton fishing boats over the breakwater as far as three blocks into town (Figure 6.4).

Other than at Kodiak, the times and sequences of wave arrivals in south-central Alaska are known only from fragmentary accounts by eyewitnesses. Most of the observers were naturally more concerned with saving their lives and property than with keeping track of changes in sea level.

Figure 6.4 Fishing boats in Kodiak, Alaska, washed ashore by the tsunami waves associated with the earthquake of 1964.

When the Warning Center at Honolulu received the first message from Kodiak, they began to prepare a tsunami warning message for dissemination across the Pacific. A more complete tide report from Kodiak came in at 0611 and a Pacific-wide tsunami warning was issued at 0637: ". . . A sea wave has been generated which is spreading across the Pacific Ocean. . . . The intensity cannot, repeat cannot, be predicted."

At 0711, with communication finally restored, Sitka, Alaska, was able to report that the tsunami had first reached there at 0510—more than 2 hours earlier.

All the while the tsunami waves continued to spread. At Chenega 23 people died when the water washed as high as a school situated 90 feet above sea level. At Port Nellie Juan, the water rose to nearly 70 feet above sea level and three lives were lost. In Orca Inlet near Cordova, the U.S. Coast Guard vessel *Sedge* ran aground when, between waves, the water level dropped 27 feet. At nearby Point Whitshed, ten cabins floated out to sea in a line like baby ducks. The owner of one of the cabins was lost—he had returned to his property thinking the tsunami was over. At the Cape St. Elias lighthouse, a member of the U. S. Coast Guard was drowned by the initial wave—and the three other men stationed there barely escaped.

There was more opportunity for a warning as time passed and the tsunami waves spread out from Alaska toward Canada. Ironically, the Canadian authorities had withdrawn from the warning system just the previous summer and there was no official warning provided to Canada. The tsunami waves struck the Canadian coast near the time of high tide, causing extensive and widespread damage. Most of this damage occurred at the twin cities of Alberni and Port Alberni, where the waves were funnelled up a narrow inlet. An alert lighthouse keeper at the mouth of the inlet who saw the arrival of the tsunami telephoned authorities in the twin cities, giving them a ten-minute warning while the waves travelled the 40 miles up the inlet. One eyewitness at Alberni reported: "I was standing in a foot of water after the first wave hit. Suddenly the second one surged up into the street. I heard people screaming and men running back and forth across the street in front of the wave. I was amazed to see two big houses, 30 feet by 50 feet and two stories high, floating out in the Somass River. They gradually broke up and sank." The water flooding into town seeped into underground storage tanks at service stations, sending gasoline flowing into the streets. Fires quickly broke out as electrical short circuits ignited the gasoline.

The highest wave reached nearly 21 feet above sea level. Although most residents were asleep when the lighthouse keeper sent his warning, there was—miraculously—no loss of life or even serious injury. The Civil Defense commander for the area later reported: "I am unable to account for the lack of casualties. . . ."

Continuing on its path of destruction, the tsunami next assaulted the coast of the state of Washington. A park ranger at Ocean City reported: "It came over the dunes shooting 5 to 6 feet high, tossing logs around like match sticks." A man stopped in the middle of the Copalis River Bridge to watch driftwood piling up against the bridge supports, when suddenly the bridge collapsed and he and his automobile plunged into the rushing water. Four boys camping at Long Beach were chased from their tent by the rising water—their car was washed away. A woman in Olympia was suddenly awakened by water rocking her trailer; she stepped out the door into waist-deep, rushing water.

At Beverly State Park on Depoe Bay near Newport, Oregon, a family of six were camping along the shore. The tsunami caught them in their sleeping bags and the four children were washed out to sea and drowned.

At Crescent City, California, the county sheriff received the alert at 11:08 P.M. local time (0708 GMT). He immediately contacted Civil Defense authorities. Low-lying coastal areas were warned and evacuated. Around midnight the first wave began to surge into town. The first two waves caused flooding and minor damage. Past tsunamis at Crescent City had been only one or two surges, which caused only minor flooding. Based on these past experiences, many people returned to their places of business in order to clean up, believing the danger to be over. Such was the case for the owner of the Long Branch Tavern and his wife, who had returned to their tavern to retrieve their money. Because everything appeared to be normal, they stopped to have a beer. Unfortunately for them, another wave struck. At 1:40 A.M. the third wave—a destructive 21-foot giant—smashed into downtown Crescent City. This wave destroyed the Long Branch Tavern, and killed the owner and his wife. Five people were drowned when the boat in which they were attempting to flee was sucked into the mouth of a creek by the wave's recession. The boat smashed against the steel grating of a nearby bridge. The third wave caused further damage by picking up a gasoline truck parked at a Texaco service station and slamming it through the garage door of a nearby automobile dealership. An electric junction box just inside the door was knocked loose by the impact, and a fire started. The fire destroyed the building and spread to the Texaco tank farm where five gasoline storage tanks exploded. The fire at the tank farm continued to burn for three days.

In all more than 300 buildings in Crescent City were destroyed or badly damaged by the tsunami. The third and fourth waves caused most of the destruction and casualties, catching those who chose not to leave, as well as those who returned to the area after the second wave.

In the San Francisco Bay area the tsunami caused extensive damage to yachts in waters near Sausalito and San Rafael. In southern California, Marina del Rey was the hardest hit. Here, 450 feet of dock was washed half a mile upchannel. At Morro Bay, as the withdrawing water tore yachts from their moorings, one observer described the scene: "It was like someone pulled a plug from the bay."

Meanwhile, tsunami waves were heading toward the Hawaiian Islands. As early as 6:25 P.M. Hawaii Civil Defense had activated its emergency operating center. At 8:43 P.M., following the announcement of a tsunami warning, the decision was made to sound coastal sirens and to activate the Emergency Broadcast System. At 9 P.M., the siren and broadcast systems were activated simultaneously in all counties. Fifteen minutes later, the Warning Center gave the ETAs for the Hawaiian Islands: Kauai, 10:15 P.M.; Oahu, 11:00 P.M.; Maui, 11:00 P.M.; Hawaii, 11:15 P.M. The sirens sounded again at 10:00 and 10:30 P.M.

The first wave report in the Hawaiian Islands came from the Coast Guard at Nawiliwili, Kauai, which recorded a 1-foot wave arriving at 10:45 P.M. Later, an 11-foot wave rolled into Kahului, Maui, and at Hilo the waves exceeded 12 and a half feet in height. In Hilo, restaurants near the head of Reeds Bay were flooded and a sidewalk at the west end of the Waiakea Bridge over the Wailoa River was undermined and collapsed. In Maui there was damage to Kahului Harbor totalling over $50,000. But there were no casualties in Hawaii, the alert and the evacuation had been successful.

At 1:00 A.M. (1100 GMT) the Warning Center issued an all-clear bulletin stating that "the larger waves have apparently passed Hawaii."

The tsunami waves continued across the Pacific but they caused no further damage. In Japan, waves only 3 to 10 inches high came ashore.

The earthquake at Prince William Sound on Good Friday 1964 resulted in 131 fatalities and between $400 and $500 million in damage. It is thought that the shallow focus of the earthquake (only 14 miles deep) contributed to the large amount of destruction. An interesting aspect of the earthquake is that the epicenter was located on land, making it the first known earthquake with a continental epicenter to be associated with a major destructive tsunami. The tsunami was generated because vertical tectonic displacement (vertical movement) took place over a large area of the seafloor.

Of the casualties, Civil Defense estimated that 122 were victims of the tsunami, whereas only 9 were victims of the earthquake itself. The death toll could have been much higher; the earthquake occurred when schools were closed and business areas uncrowded. Moreover, if the tsunami waves had struck the coastal communities of south-central Alaska at high tide rather than low tide, the loss of life and property would have been even greater.

California was, perhaps, the only place where the number of deaths could have been much smaller. Eleven people were killed and 35 injured in Crescent City, and total damage was assessed at approximately $10 million. There was adequate warning and time for preparation, yet the death toll and amount of damage in California was greater than any other place except Alaska. The problem lay partially in the poor response to the alert and the lack of public understanding of tsunamis.

Reaction by county and city Civil Defense organizations varied considerably. In Humboldt County, the second advisory issued by the Warning Center was received at 11:08 P.M. local time at the county sheriff's office. All agencies were mobilized by 11:18 P.M., and evacuation of all persons in danger areas was completed and road blocks established by 11:40 P.M.

Immediately upon receipt of the advisory in San Francisco, attempts were made to evacuate coastal areas. Yet an estimated 10,000 people jammed the beaches to watch the tsunami waves arrive. In San Diego, attempts to evacuate the beaches were rendered useless by curious onlookers. Los Angeles County made no attempt whatsoever to evacuate the waterfront. If large waves had struck these areas the casualty lists could have been staggering.

Why were the waves associated with this giant earthquake so small in Hawaii? Experts believe that the geometry of the fault caused maximum energy to be projected perpendicular and not parallel to the fault. Because the northwest coast of North America is situated at right angles to the fault, it received the largest tsunami waves. Luckily, the Hawaiian Islands were not in the direction of maximum energy propagation.

But Alaska had not been so fortunate, and a glaring gap was revealed in the warning system—there had been no warning procedure for locally generated tsunamis. As a result of the 1964 earthquake, the Alaska Regional Tsunami Warning System (ARTWS) was established in 1967 in order to provide timely tsunami watch and warning to Alaska for locally generated tsunamis.

At its inception the ARTWS consisted of a main observatory at Palmer—40 miles north of Anchorage, and two secondary observatories at Sitka and Adak. The system has since been renamed the Alaska Tsunami Warning Center (ATWC), and only the headquarters at Palmer is still in operation.

Because of the speed with which a locally generated tsunami can strike the various areas of the Alaskan region, the ATWC must react very quickly to all earthquakes that might possibly generate tsunamis. As a result, all personnel are required to live and remain within five minutes travel time to the center. The staff is ready to respond to alerts of large earthquake events by a special radio-alarm system (RAS) composed of alarms installed in staff residences and VHF (very high frequency) pocket voice receivers carried by staff at all times.

An earthquake of Richter magnitude 6.0 or greater within a radius of about 600 miles activates the RAS and initiates an earthquake/tsunami investigation. The investigation begins by locating the epicenter and determining the size of the earthquake and culminates in either processing the event routinely or the issuance of a combined tsunami watch/warning. A watch/warning is automatically issued when the magnitude of an earthquake in Alaska is 6.75 or greater and occurs near a coastal area.

Those areas within three hours wave travel-time from the earthquake epicenter are immediately placed on a warning status (without confirmation, owing to the lack of time). Areas outside the warning area are placed on a watch status.

Next the tide stations nearest the epicenter are monitored for the existence of a tsunami. If a tsunami has been generated, the watch areas are upgraded to a warning status. The warning and other emergency information are sent out by the Emergency Broadcast System, VHF radio, commercial telephone lines, Coast Guard and Marine

Weather HF radio, the National and Alaska Warning Systems, and military communication channels.

As experience in Hawaii and California has shown, it is not enough simply to warn the population. The public must understand the warning and respond. With this goal in mind, the ATWC maintains a community preparedness program designed to educate the public on what to do in the event of a violent earthquake or tsunami. The program goes to each community and presents a detailed briefing covering the seismicity of their area, historical earthquake/tsunami events and damage, and a description of what might happen if an earthquake or tsunami were to occur.

During the past 17 years, the ATWC has responded to an average of more than a dozen alerts each month. The Alaska Tsunami Warning Center has done an excellent job of informing residents of the seismically active Alaska region about the dangers from local earthquakes and tsunamis.

Following the 1960 Chilean tsunami, the Pacific Tsunami Warning Center in Honolulu expanded as new members joined the system. An international coordination group was established to review the activities of the Warning System, and in 1965 the International Tsunami Information Center began working under the auspices of UNESCO (United Nations Educational, Scientific, and Cultural Organization).

With the headquarters of the improved and expanded Warning System situated in Honolulu, would there ever be a need for a special local warning system for the Hawaiian Islands? Could Hawaii be in danger from a locally generated tsunami?

CHAPTER SEVEN

Local Tsunamis in Hawaii

As described in earlier chapters, most destructive tsunamis are associated with earthquakes and are caused by tectonic displacement of the seafloor. Fortunately, most of the Hawaiian Islands, with the exception of the island of Hawaii itself, are not very seismically active. The island of Hawaii does have a large number of earthquakes, but most of these are small and cause little or no damage. About once a century, however, a very large earthquake does occur on the island. Such a quake occurred in 1868.

The Great Earthquake of 1868

On Thursday, April 2, 1868, a major earthquake struck the island of Hawaii. The quake was felt as far away as Kauai, but caused extensive damage only on the Big Island, where it was reported that every European-style building in the Kau district (Figure 7.1) was completely destroyed.

An eyewitness to the destruction related his experience (Cox and Morgan 1977):

> At 4 P.M. on the 2nd a shock occurred, which was absolutely terrific. All over Kau and Hilo, the earth was rent in a thousand places, opening cracks and fissures from an inch to many feet in width, throwing over stone walls, prostrating trees, breaking down banks and precipices, demolishing nearly all stone churches and dwellings, and filling the people with consternation. This shock lasted about 3 minutes, and had it continued three minutes more, with such violence, few homes would have been left standing in Hilo or Kau. Fortunately there was but one stone building in Hilo, our prison, and that fell immediately.

Figure 7.1 Map of the island of Hawaii showing selected locations mentioned in stories of the 1868 and 1975 earthquakes and tsunamis.

The earthquake, which probably had a Richter magnitude between 7.25 and 7.75, triggered a large landslide in Wood Valley on the slopes of Mauna Loa, about 5 miles north of Pahala. Here a large mass of lava rock slid downslope over a layer of wet ash, going from an altitude of 3,500 feet down to 1,620 feet. In all, the slide covered a distance of some 2 and a half miles and completely destroyed a Hawaiian village. Thirty-one natives, and more than 500 horses, cattle, and goats were buried alive.

As if the earthquake weren't enough, a tsunami was generated! A Mr. Stackpole, returning from the Volcano House at Kilauea to the shore at Keauhou, met the men who worked at the Keauhou landing running up hill. They reported that immediately after the earthquake the sea had rushed in and "swept off every dwelling and store house . . . " (Dana 1868).

According to another account, the tsunami "rolled in over the tops of the coconut trees, probably sixty feet high, and drove the floating rubbish, timber, and so forth, inland a distance of a quarter of a mile in some places, taking out to sea when it returned, houses, men, women, and almost everything movable. The villages of Punaluu, Ninole, Kawaa, and Honuapo were utterly annihilated (Whitney 1868).

It is now estimated that the tsunami waves had a run-up height of about 45 feet at the Keauhou landing and 9 feet at Hilo. This was definitely a destructive local tsunami and resulted in 46 deaths in Hawaii. The waves spread out across the Pacific, registering on tide gauges in Oregon and California some five hours later. But outside of Hawaii, the waves were very small—measuring only 4 inches at San Diego—and caused no damage.

The 1868 tsunami produced one of the most fantastic of all stories about tsunamis. The tale, as related by Mr. C. C. Bennett (1869), follows:

> I have just been told of an incident that occurred at Ninole, during the inundation of that place. At the time of the shock on Thursday, a man named Holoua, and his wife, ran out of the house and started for the hills above, but remembering the money he had in the house, the man left his wife and returned to bring it away. Just as he had entered the house the sea broke on the shore, and enveloping the building; first washing it several yards inland, and then, as the wave receded, swept it off to sea, with him in it. Being a powerful man, and one of the most expert swimmers in that region, he succeeded in wrenching off a board or a rafter, and with this as a *papa hee nalu* (surfboard) he boldly struck out for the shore, and landed safely with the return wave. When we consider the prodigious height of the breaker on which he rode to the shore, (50, perhaps 60 feet), the feat seems almost incredible, were it not that he is now alive to attest it, as well as the people on the hill side who saw him.

Both the epicenter of the 1868 earthquake and the site of generation of the tsunami lay near Kalapana, off the southeast coast of the island of Hawaii. Nearly a century would pass before this area would again produce a large earthquake. On November 29, 1975, such an earthquake would occur.

The Halape Tragedy

Thanksgiving weekend. Time for a break. What better way to spend it than to go camping at the remote beach park of Halape on the island of Hawaii (Figure 7.2). One of Hawaii's most idyllic retreats, Halape lay just seaward of the base of the 1,000-foot cliffs of Puu Kapukapu (Forbidden Hill). Because it could be reached only by foot or on horseback, Halape remained an unspoiled spot in paradise. Thirty-four people, including several fishermen, a group from the Sierra Club, and Boy Scout Troop 77, decided to enjoy the special pleasures offered by Halape on that holiday weekend in 1975.

The campers from Troop 77 included six Scouts accompanied by four adults. They hiked in on Thanksgiving Day and settled down to have a good time. The four men had been looking forward to the trip even more than the boys. Dr. James Mitchell had arranged his schedule so that he could go with Claude Moore and Don White, and Policeman James Kawakami had flown in from Honolulu that morning, directly after finishing his night shift. When they arrived at the beach, the four men set up camp in one shelter while the boys selected a spot in the coconut grove.

It rained that first night, so the boys—David White (Don's son), "Fal" Allen, Mike Sterns, Leif Thompson, Noel Loo, and Timothy Twigg-Smith—moved to a second shelter about a quarter of a mile along the beach. The rain also created somewhat of a problem for the

Figure 7.2 Aerial photo of the Halape palm grove prior to the 1975 tsunami and earthquake.

adults, as it caused the water tank to overflow into the shelter where they were sleeping. The water soaked their shoes, and so they hung them from the roof to dry.

The next day was spent fishing, swimming, and being lazy. That night they went to sleep, situated as they had been the night before, except that David White decided to pitch his tent in the coconut grove.

At 3:36 A.M. the island was jolted by an earthquake. The quake, which measured 5.7 on the Richter scale, was centered beneath the south flank of Kilauea Volcano, about 3 miles inland of Kamoamoa.

At Halape the campers were startled awake by the movement of the earth. Hearing the crashing sound of rock falls, they looked toward Puu Kapukapu. In the weak early-morning light, they saw dust clouds rising as rocks fell down the steep cliff face.

After chattering about the quake for a while, some of the campers went back to their shelters to sleep. Others, however, decided to avoid the danger of landslides, and moved closer to the ocean. What they failed to realize was that the main danger was to come from the sea. David told his father that he felt something was going to happen and that he wanted to leave. He was reassured and went back to sleep.

Little more than an hour later, at 4:48 A.M., a second earthquake struck. This quake was also beneath Kilauea's south flank, but it was centered southeast of the Wahaula *heiau,* about 2 miles offshore, and unlike the earlier quake, it did not subside in intensity. The campers on the beach at Halape experienced a terrible shaking and shuddering. Many parts of the island were plunged into darkness as the violent trembling felled utility poles and caused power outages. The quake was so severe locally that the seismographs at the Hawaiian Volcanoes Observatory went off scale! A Richter magnitude of 7.2 was later calculated by scientists at distant stations in North America, Japan, and New Zealand, where the seismographs had remained on scale. The last earthquake of this magnitude in Hawaii had been the great earthquake of 1868, and it was accompanied by a tsunami!

At Halape, only 15 miles west of the epicenter, Claude Moore and his companions woke to find the shelter shaking and the water tank rocking and splashing water. Thinking that the water tank might collapse on them, they ran outside. Although able to stand when the quake first started, they were now thrown to the ground by the violent shaking. Next they heard a terrifying roar as boulders crashed down the cliff face of Puu Kapukapu. Some of the campers began scrambling toward the beach, away from the threat of falling rocks. Claude ran back into the shelter and began fumbling with the laces of his shoes where they had been tied to the roof to dry. The other adults were outside talking with two campers who had just been drenched by a wave as they slept on the beach. Claude heard someone say something about water and thought of the possibility of a tsunami. All of a sudden there was silence. He ran out of the shelter toward the beach to see what was happening.

Claude looked to the right and saw a group of people running as fast as they could away from the beach toward Puu Kapukapu. He looked back to his left and saw a 5-foot wall of water crashing through the coconut palms. It was almost upon him. He flung himself behind the 4-foot stone wall at the back of the shelter and hung on for all he was worth. At first he was doing well, holding on to the top of the wall and floating on the surface of the water. Then the pressure of the wave proved too much for the stones and the wall started to collapse. At the same time the shelter was swaying and creaking, so he let go. "This is it," he thought as he was washed back and forth in the water. It seemed that he was swirled in the water for an eternity—although it may have been only a few minutes. Suddenly he was on his feet, the water was waist-high. A ray of hope penetrated the mist of his confusion. Perhaps there was a possibility of escape after all!

A great blow from behind drove all thoughts from Claude's mind. The second wave, some 25 feet high, overwhelmed him. His face crashed onto a rock, stars flashed behind his eyelids, and the water was over and around him—then it was gone.

Claude found himself at the base of a crevice, some 20 feet deep and 30 feet across. He clambered up the back of the crack, realizing with a sense of wonder that he was still alive. Battered and abraded, his shirt hanging off his arms by the cuffs and his watch torn from his wrist by the wrenching wave, he was dazed and bewildered—but he was alive! He had no serious injuries, and began to wonder about the others. He stood up and tried to walk but became sick and dizzy. Once he heard Don White calling for David, and called back to him. After that he heard no sound but the sea.

Just as the sea had destroyed the shelter to which Claude had been clinging, so it ravaged the structure where the Boy Scouts were sleeping. The arrival of the wave was the first they knew of their danger. The rushing water tore the shelter apart, pushing Timothy through the wall. All the boys became entangled in bushes and pulled to and fro by the water, until the waves receded and they were able to run to higher ground. There they waited until daybreak with three fishermen from Keaau.

The group of fishermen had ridden into Halape on horseback and camped near the animal corral at the back of the beach. After they were wakened by the second earthquake they heard sounds of a landslide coming from the cliffs, and they ran toward the sea. They immediately reversed their direction, however, when they heard someone shouting "Tidal wave!" They ran before the advancing wave together with several of the members of the Boy Scout party. They ran until they were confronted by a deep ditch, 8 to 10 feet wide. There was no choice, nowhere else to go. One of the campers, Michael Cruz, hesitated for a moment and disappeared forever into the sea. The rest jumped into the ditch, and the wave followed, crashing on top of them. Whenever they tried to climb out, more water would pour onto them, tumbling them over and around. According to one survivor, it was like being "inside a washing machine." The second wave, higher and more turbulent than the first, had washed everything in its path as far as 300 feet inland. Trees, debris from the shelters, rocks, and people were deposited in the ditch. Several smaller waves washed over the exhausted victims, but their uncomfortable refuge saved them from the fate of being carried out to sea. All those who had been in the ditch survived, although James Kawakami and Don White had swallowed a great quantity of seawater and sand. Both Don and David had been pounded by the water and debris, but David was not hurt.

After the waves had receded, David took care of his father. He pulled Don into the shelter of a rock, covering his body with long grass and then lying over him to provide his own body warmth.

When day dawned, Lief Thompson, who was not hurt, went searching for the others. He found Claude and reported to him that all members of the Boy Scout party were safe except for Dr. Mitchell, but that Jimmy Kawakami was in a state of shock after his ordeal in the ditch. Lief returned to Jimmy and stayed with him until help arrived.

Both Lief Thompson and David White received awards from the Boy Scouts of America for their life-saving efforts during the disaster.

All the campers were rescued later that day by army helicopter, and taken to Hilo Hospital. Nineteen of the victims had been injured, seven requiring extended hospitalization. Four of the ten horses were lost. Michael Cruz had been taken by the sea and his body was never found. Of Claude's party, all were safe except for Dr. Mitchell. He had been battered and drowned among the rocks and debris of Halape.

~

The locally generated tsunami that struck Halape was caused by the sudden movement of the seafloor off the southeast coast of Hawaii, the same movement that produced the 7.2 earthquake. Coincident with the earthquake, the ground along the shoreline subsided by up to 10 feet, submerging much of the Halape palm grove (Figure 7.3). It is thought that the first wave was caused by water along the shoreline rushing in to fill the cavity caused by the sudden subsidence. The second wave, which was much larger, resulted as the deep-water tsunami wave arrived from offshore.

Amazingly, the loss of life was confined to Halape, but property damage was widespread. The tsunami had radiated out in all directions from its source, travelling several hundred miles per hour.

At the small bay of Punaluu, 20 miles southwest of Halape, residents in their homes as well as several families camping near the beach were awakened by the earthquakes. What no one suspected was that a tsunami was headed toward them. Many residents and campers were forced to wade to higher ground as sea level rose steadily and quickly. But the big wave would not arrive for another ten minutes.

Cecil S. Carmichael and his wife were wakened by the foreshock at 3:38 A.M., but remained in their A-frame house in Punaluu. After the big quake at 4:48, the Carmichaels got dressed and decided to go downstairs and drive away to higher ground. Their account of the tsunami is as follows (Cox and Morgan 1977):

> Before we could leave the house, we heard a sound like a strong wind. My wife pulled back the front drape and saw water over our deck and coming in under the front sliding door. This was at 5:00 A.M. Our lot is +9 ft. and the house was 4 ft. off the ground; i.e. the floor level was at +13 feet.
>
> We left by the back door using flashlights. The back steps were gone—also our 3000-gal. water tank and garage. We waded through knee-deep muddy water and climbed over wet stone walls and walked quickly in 1-ft.-deep water to the main paved road, going through Dahlberg's yard. There the lei stands had been smashed back about 30 ft. and a power line

Figure 7.3 Aerial photo of the Halape palm grove after the 1975 tsunami and earthquake.

was down on the pavement. We heard another sound like a strong wind. My wife ran, and I shone my light and saw the second wave by the old wharf. This was at 5:10 A.M. I ran. When I reached the new road, I heard the crashing of metal and wood. The wave reached the lawn of Arnold Howard's house. I later measured the wave height at two coconut trees and two hala trees in our lot using a tape, a staff, and a hand level. The height was 25 feet above sea level.

The third wave which hit at 5:20 A.M., must have had a height of at least 13 feet but less than 25 feet.

Our house dragged a coconut tree with it 100 feet inland and broke apart—a total loss. We could see most of it afterward floating offshore with other debris; and fishermen friends have told us they subsequently found lumber from it washed up at several locations between Punaluu and South Point. Our four stone walls were completely washed away. Our car by the wall was turned around 180° and was a total loss.

Throughout Punaluu, beach houses and ocean-front properties were swept from their foundations (Figure 7.4), and the Punaluu Village restaurant and gift shop were inundated. Although the restaurant structure remained intact, interior damage would amount to

Figure 7.4 House at Punaluu washed off foundation by the tsunami of 1975.

nearly $1 million; the waves left the floor covered in mud and the furniture jumbled and broken. Throughout the area, evidence of the advance of the sea could be seen in the form of stranded sea urchins, eels, fish, and starfish. In all, seven homes and two vehicles were destroyed by the tsunami but no injuries were reported.

Farther down its path of destruction, the tsunami claimed park facilities, a warehouse, and a fishing pier at Honuapo. At Kaalualu Bay, 15 miles southwest of Punaluu, the waves damaged several vehicles and ravaged a campsite, terrifying seven campers.

Although near the epicenter of the earthquake the waves were from 20 to nearly 50 feet high, they diminished in height rapidly as they spread away from their source.

~

The big earthquake had shaken all of Hilo. Some buildings had been structurally damaged, chimneys toppled, windows were broken, and merchandise was knocked from store shelves. The edge of the parking lot at the Bayshore Towers condominium had fallen into the bay. Five miles north of Hilo, the scenic Hoaloa Arch had collapsed during the quake.

The first tsunami wave reached Hilo twenty minutes after the big quake. Although this wave was only 1 and a half feet high, it was followed by a major recession of the water some 5 feet below normal. At tiny Radio Bay, located in the corner of Hilo Harbor, the crew members of the Coast Guard cutter *Cape Small* watched helplessly as their ship settled into the mud and began to list to one side. Then the sea surged back in as the second and largest wave crested in Hilo Bay at about 5:30 A.M. This wave, up to 8 feet high, ripped small boats from their moorings, washing some onto the pier and sinking others. An automobile was swept off the pier into the harbor. In one peculiar incident, a man was thrown from his boat onto the pier by the advancing wave, then washed off the pier into the water, and back onto his boat as the wave receded.

A series of progressively smaller waves continued to surge in and out of Hilo Bay at approximately 15-minute intervals for several hours.

The tsunami continued to spread out across the Pacific. The waves arriving in Honolulu 30 minutes later were about a foot high. The tsunami measured 1 foot in Los Angeles, where it was recorded 6 hours and 45 minutes after the earthquake, having crossed the Pacific from Hawaii at a speed over 350 miles per hour. The waves took some eight hours to reach Japan, where the largest wave measured almost 2 feet. In all, at least 39 tide stations outside of Hawaii recorded the tsunami.

Death and destruction from the tsunami occurred only on the island of Hawaii. The run-up height of the largest wave was 47 feet at Keauhou Landing, 26 feet at Halape, 25 feet at Punaluu, and 8 feet at the Wailuku River in Hilo Bay.

On the island of Hawaii, damage from the earthquake and tsunami exceeded $4 million. A total of 8 houses, 3 business, and 27 fishing boats were destroyed or severely damaged by the tsunami. Two people were killed, both at Halape. The President of the United States declared the County of Hawaii a disaster area.

How had the authorities reacted to this locally generated tsunami? Had there been time for a warning?

The first reaction to the tsunami occurred when reports of unusual wave activity in Kona reached the Hawaii County Police. They immediately ordered the evacuation of the Hilo waterfront. State Civil Defense ordered the coastal sirens to be sounded on Hawaii and Maui. But the first wave arrived in Hilo almost 30 minutes before the sirens sounded. Coastal sirens on Maui were not sounded until 7:20 A.M., over 1 and a half hours after the earthquake. The alert was cancelled at 8:30 A.M.

The implications are clear. Had the tsunami occurred at a time other than in the early morning hours when few people were at the harbor or along the shore, or had the waves been larger, the death toll could have been much greater.

The two most important local tsunamis in the history of Hawaii were associated with Hawaii's two largest earthquakes—those of 1868 and 1975. Both tsunamis are thought to have been generated by movement on underwater portions of the southeast flank of Kilauea volcano. Kilauea continues to spout lava and build the island edifice: no doubt the southeast flank will again produce a major earthquake and perhaps a tsunami.

Local Volcanic Tsunamis

Has movement of the seafloor been the only source of locally generated tsunamis in Hawaii? Could Hawaiian volcanoes produce an explosion like Krakatoa and generate a giant tsunami? Fortunately, this is extremely unlikely. Volcanoes of the type found in Hawaii tend to be effusive and not explosive. A tsunami generated by a volcanic explosion has never been known to occur in Hawaii. However, one important local tsunami was related to volcanic activity.

This volcanic-related tsunami occurred along the Kona coast of the island of Hawaii on October 2, 1919. On September 26 of that year, Mauna Loa volcano began to erupt from its summit caldera. At about midnight on September 28, several vents opened on the south rift and lava began to flow toward the sea. At about 4:30 Saturday afternoon, September 30, the lava reached the shore at Alika on the Kona coast. Large volumes of lava poured into the water, and, after three days, the flow had built a delta out into the sea.

People from all over the island travelled to Kona to witness the phenomenon. Among these were the Carlsmith family. They had driven from Hilo to arrive at Hoopuloa at about midnight on October 1. At 7:30 the next morning, they went down to the Hoopuloa wharf to take a canoe out to watch the lava flowing into the sea. Mr. Carlsmith's account was quoted in the *Hilo Daily Tribune* (3 October 1919):

> My first intimation that a tidal wave was impending was the recession of the sea. Suddenly it seemed to slope backward from the shore. In a moment it was visibly running downhill. The rugged rocks of the coastline were exposed and stranded fish were left flopping on the shore.
>
> The others about me did not seem to realize what was coming. My own recollections are rather confused, but I remember that I knew we were to have a tidal wave.
>
> Then the water came rushing back. I should say they [the waves] were 12 to 14 feet higher than the high water mark.
>
> A man who has been through such an experience cannot presume to speak with precision, but I should say there were at least 10 separate waves, the first was the strongest. Two others of nearly equal volume followed it, then there was a period of agitated intermission and the other waves swept inward.

The tsunami is thought to have consisted of a series of waves, 3 to 14 feet high, following quickly one behind the other. Although the waves were probably largest near the delta of the lava flow, they managed to flood the Hoopuloa wharf at the shore.

Carlsmith and his wife and two sons were standing on the wharf when it was engulfed by the waves. One of the sons ran toward their automobile to try and save it, but was forced to abandon it as it was wrecked by the surging water.

The other son was washed into the wharf shed, where he managed to grab a beam and hang on as the wave receded. Carlsmith himself was washed off the pier and carried about a hundred yards offshore. He struggled through the turbulence and managed to regain the land. Here he found his sons safe, but his wife was gone. She had been carried almost a quarter of a mile out to sea, where she could be seen struggling to remain afloat. Frantically, Carlsmith searched for a canoe, but those on shore had all been wrecked by the waves. Responding to his pleas for help, two fearless Hawaiian men swam out to a canoe floating offshore and then paddled out to rescue Mrs. Carlsmith.

Following the tsunami, the *Hilo Daily Tribune* (4 October 1919) reported: "It is now nearly impossible to get a Hawaiian boatman to take spectators to view the cascading of the lava into the sea by boat from Hoopuloa. Frankly, they admit fear of another tidal wave, although many believe the real reason is superstition regarding the wrath of Pele."

The exact cause of this tsunami is not definitely known, but the

Hawaiian Volcanoes Observatory registered no unusual earthquake activity during the period. One possibility is that as the lava continued to flow into the sea the front of the delta became increasingly steep until part of it collapsed, slumping down slope. Such an underwater slump or landslide would have gone unobserved but could have generated a small local tsunami.

The 1919 tsunami, although ultimately a result of volcanic activity, may really have been caused by a submarine landslide. Landslides, as you will recall, are one of the potential causes of local tsunamis. Have there been other landslide-generated tsunamis in Hawaii?

Tsunamis Caused by Landslides

In the historic records of tsunamis in Hawaii there is only one tsunami that can definitely be attributed to a landslide. This occurred at Napoopoo on August 21, 1951, and there was not one but two separate local tsunamis that day.

Just before 1 A.M. a strong earthquake, centered about 3 miles west of Napoopoo, shook the Kona coast of the island of Hawaii. The quake, measuring 7.0 on the Richter scale, was caused by movement along the Kealakekua fault that passes along the north coast of Kealakekua Bay (opposite Napoopoo) and extends westward out to sea.

At Napoopoo the sea began to withdraw. Fearing a tsunami, most of the villagers fled to higher ground. The water level dropped 4 feet and then rose 2 feet above normal. The tsunami was too small to do any damage. Farther down the coast at the village of Milolii, the water rose 4 feet and floated a canoe off the beach. The small fault-generated tsunami was recorded on the tide gauges in both Hilo and Honolulu.

But this was to be only the first tsunami of the morning. About two hours later, another small tsunami struck Napoopoo. According to local police, a large piece of the steep cliff forming the northern shore of Kealakakua Bay plunged into the water. The enormous splash sent 2-foot waves ashore at Napoopoo.

If 2-foot tsunami waves don't seem too impressive, scientists of the U.S. Geological Survey have recently documented an ancient landslide-generated tsunami of truly gigantic dimensions (Moore and Moore 1984). According to the geologists, the origin of a deposit of reef debris and rock found more than 1,000 feet above sea level on

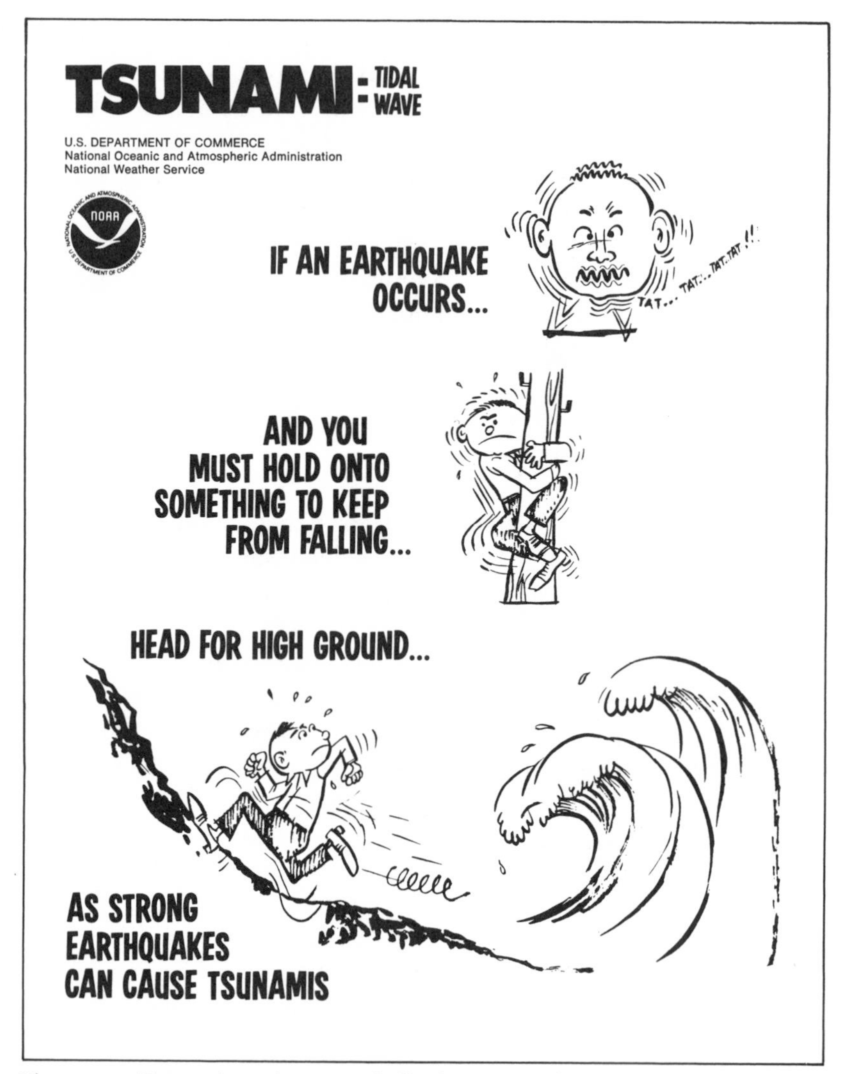

Figure 7.5 Tsunami warning poster indicating what to do when a strong earthquake occurs in Hawaii.

the island of Lanai can probably be attributed to a giant wave "that swept several hundred meters up the flanks of Lanai and nearby islands" and deposited the material before it receded. They speculate that the wave was caused by an underwater landslide along the Hawaiian Ridge near Lanai. This cataclysm occurred about 100,000 years ago!

The chances of another such event are considered remote. And with large earthquakes occurring only about once a century, do residents of Hawaii really need to be concerned about local tsunamis?

The answer is an unqualified "yes!" Since 1848, there has been no period longer than 35 years without a local tsunami in the Hawaiian Islands.

Because of the very short interval between the occurrence of an earthquake in Hawaii and the arrival of tsunami waves in populated areas, it may not always be possible to issue a tsunami warning in time. What can be done to avoid being caught in a locally generated tsunami? The Tsunami Warning System advises that people in coastal areas should immediately evacuate anytime a strong earthquake strikes (Figure 7.5). In other words, if the ground really shakes, head for the hills!

CHAPTER EIGHT

The Next Tsunami?

THE THREAT to the Hawaiian Islands from both locally generated and Pacific-wide tsunamis remains real. This doesn't mean that residents and visitors should live in a constant state of fear, but rather they should understand the meaning of the warnings and know what action to take. Unfortunately, the support for a public education campaign about tsunamis has not been forthcoming. This remains one of the priorities and goals of both the Warning System and Hawaii Civil Defense agencies.

The Tsunami Warning System (TWS) of the Pacific has been improved and expanded. A large number of countries and territories joined soon after the destructive Chilean tsunami. Hawaii is no longer without warning of tsunamis coming from the south. There are now many tide stations between South America and Hawaii, including Kanton (Kiribati), the Johnston Islands (U.S.A.), Baltra Island (Ecuador), and Isla de Pascua and San Felix Island (Chile). In fact, Chile now maintains its own reliable local tsunami warning system.

Today, the TWS consists of 62 tide stations, 77 seismic stations, and hundreds of points for the dissemination of information (Figure 8.1) scattered throughout 24 member countries in the Pacific Basin. Headquarters of the TWS is at the Honolulu Observatory, now called the Pacific Tsunami Warning Center (PTWC), operated by the National Weather Service of the U.S. National Oceanic and Atmospheric Administration (NOAA).

The transmission of seismic data to the PTWC, as well as research on earthquakes, is a coordinated effort of the National Weather Service and the U.S. Geological Survey. Tsunami research is carried out

Figure 8.1 *(opposite)* Map of seismograph and tide reporting stations of the Tsunami Warning System.

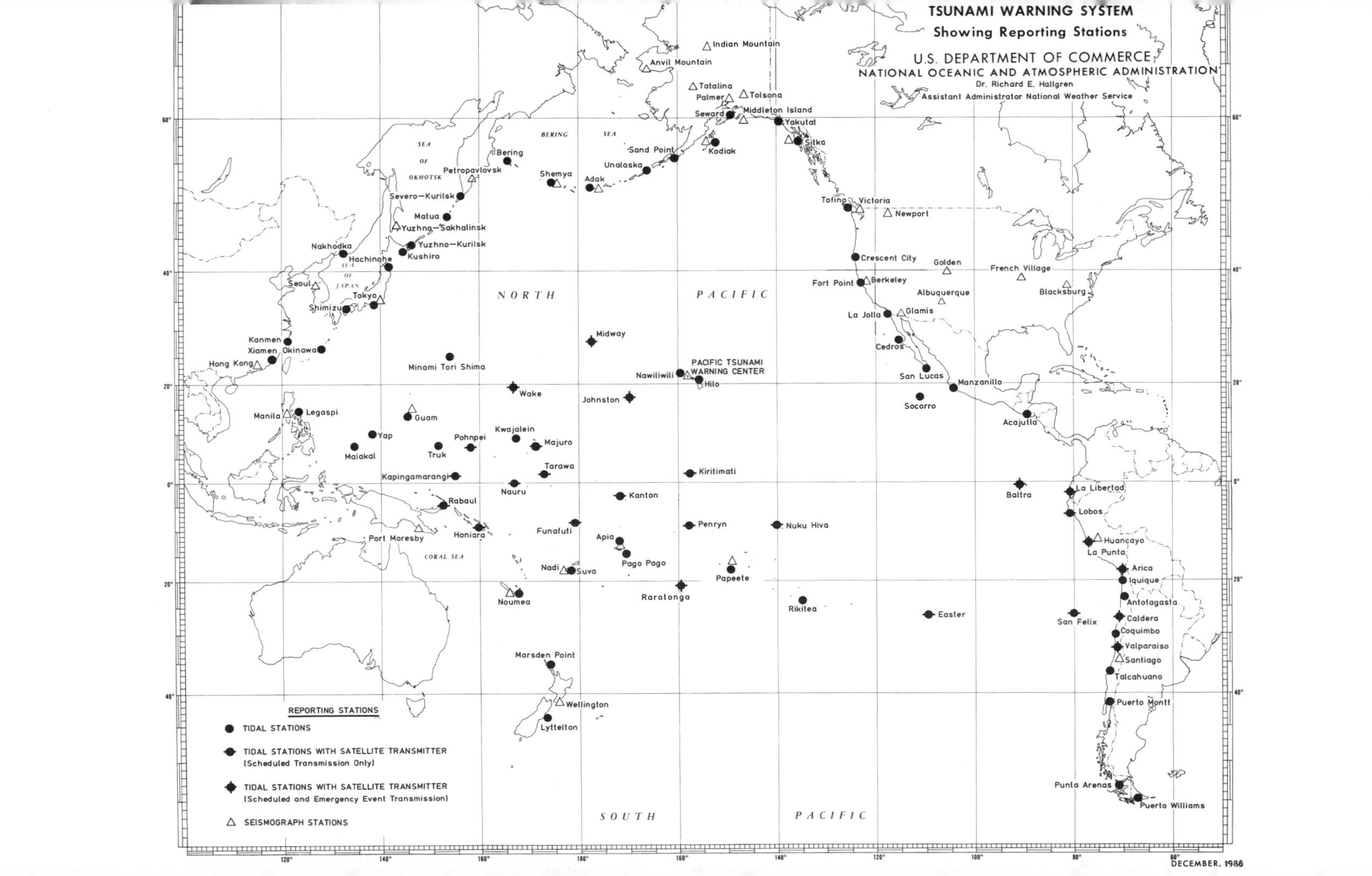
TSUNAMI WARNING SYSTEM
Showing Reporting Stations
U.S. DEPARTMENT OF COMMERCE
NATIONAL OCEANIC AND ATMOSPHERIC ADMINISTRATION
Dr. Richard E. Hallgren
Assistant Administrator National Weather Service
REPORTING STATIONS
TIDAL STATIONS
TIDAL STATIONS WITH SATELLITE TRANSMITTER
(Scheduled Transmission Only)
TIDAL STATIONS WITH SATELLITE TRANSMITTER
(Scheduled and Emergency Event Transmission)
SEISMOGRAPH STATIONS
NORTH PACIFIC
SOUTH PACIFIC
SEA OF OKHOTSK
BERING SEA
SEA OF JAPAN
CORAL SEA
Indian Mountain
Anvil Mountain
Tatalina
Palmer
Tolsona
Seward
Middleton Island
Yakutat
Sitka
Kodiak
Sand Point
Unalaska
Adak
Shemya
Bering
Petropavlovsk
Severo—Kurilsk
Matua
Yuzhno—Sakhalinsk
Yuzhno—Kurilsk
Kushiro
Nakhodka
Hachinohe
Seoul
Tokyo
Shimizu
Kanmen
Xiamen
Okinawa
Hong Kong
Minami Tori Shima
Midway
PACIFIC TSUNAMI WARNING CENTER
Nawiliwili
Hilo
Johnston
Wake
Manila
Legaspi
Guam
Yap
Malakal
Truk
Pohnpei
Kwajalein
Majuro
Tarawa
Kapingamarangi
Nauru
Kiritimati
Kanton
Rabaul
Honiara
Port Moresby
Funafuti
Apia
Pago Pago
Penryn
Nuku Hiva
Nadi
Suva
Noumea
Rarotonga
Papeete
Rikitea
Easter
Marsden Point
Wellington
Lyttelton
Tofino
Victoria
Newport
Crescent City
Golden
French Village
Fort Point
Berkeley
Albuquerque
Blacksburg
La Jolla
Glamis
Cedros
San Lucas
Manzanillo
Socorro
Acajutla
Baltra
La Libertad
Lobos
Huancayo
La Punta
Arica
Iquique
Antofagasta
San Felix
Caldera
Coquimbo
Valparaiso
Santiago
Talcahuano
Puerto Montt
Punta Arenas
Puerto Williams
60°
40°
20°
0°
20°
40°
120°
140°
160°
180°
160°
140°
120°
100°
80°
60°
DECEMBER, 1986

by NOAA's Environmental Research Laboratories, the Joint Institute for Marine and Atmospheric Research at the University of Hawaii, and by various other universities under the direction of the National Science Foundation.

Since the last major Pacific-wide tsunami in 1960, a number of advances have been made in our knowledge of the behavior of tsunamis. These advances have come from research based on actual observations of the waves and from experiments using tsunami models.

Tsunami Monitoring

The best way to learn about tsunamis is, of course, to study the tsunami waves themselves. Because tsunamis are not everyday events, it is imperative to collect the most information possible from each occurrence. This task has been coordinated under the Tsunami Monitoring Program directed by the University of Hawaii.

When a Tsunami Watch is declared, trained volunteer observers, professional scientists, and cooperating military personnel begin preparing their equipment. By the time a Tsunami Warning goes into effect, they are ready to put the Monitoring Program into operation.

Observers head toward preselected shoreline vantage points. Here, 8 millimeter timelapse surveillance cameras are set up to film the waves. At sites safe from the advancing tsunami, observers man the equipment and note the wave activity. At less-secure sites the cameras are set to run automatically after the area is evacuated.

Recently developed portable tsunami gauges (Figure 8.2) are deployed from piers and in designated shoreline areas. These gauges sense the change in water pressure as the waves pass over them and record their measurements in a semi-conductor memory (by computer microchip).

U.S. Navy P-3B patrol planes take to the air prior to the arrival of the first waves. The aircraft fly at an altitude of 1,000 feet in an elliptical pattern over critical shoreline areas, covering the same spot every 15 minutes. Special cameras mounted in the belly of the plane take 480 exposures on 70 millimeter film, documenting the arrival of each tsunami wave on the shores of the Hawaiian Islands.

After the tsunami waves have passed and it is once again safe to enter shoreline areas, the portable tsunami gauges and surveillance cameras are recovered.

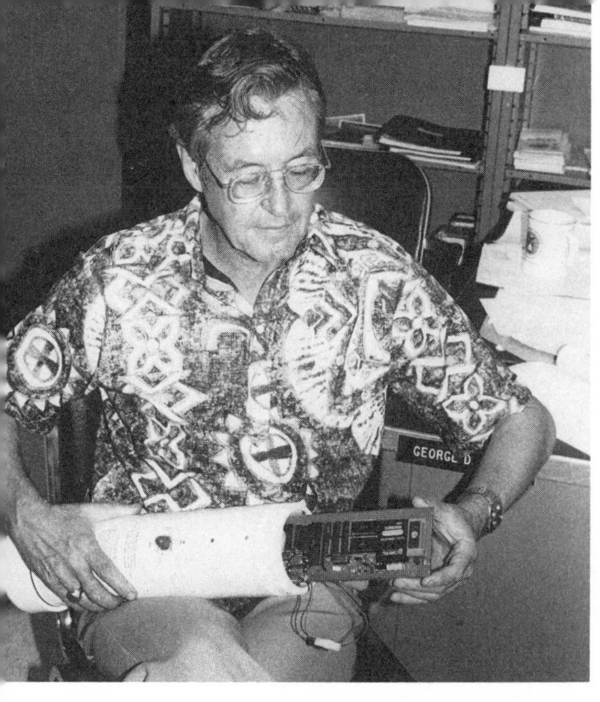

Figure 8.2 George Curtis of the Joint Institute for Marine and Atmospheric Research of the University of Hawaii displays a prototype portable gauge for measuring tsunami waves.

A ground survey and damage assessment is conducted by a team from the U. S. Army Corps of Engineers, joined by volunteers from the American Society of Civil Engineers.

A post-tsunami aerial photographic survey is undertaken jointly by the U.S. Navy, the Coast Guard, and the Civil Air Patrol, and even makes use of special National Weather Service U-2 surveillance aircraft.

To remain in a state of readiness, the Tsunami Monitoring Program holds yearly meetings and conducts rehearsals of most aspects of the program.

Tsunami Modeling

The damage caused by destructive tsunamis is a result of the run-up of the giant waves on shore. Various modeling techniques have been used to try to simulate the run-up phase. One such technique is called hydraulic modeling and uses a physical scale-model for experiments.

Because of Hilo Bay's peculiar sensitivity to tsunami waves, a hydraulic model of the bay was constructed following the 1960 tsunami (Figure 8.3). The model measures 85 feet by 62 feet and represents the triangular shape of Hilo Bay. Model tsunami waves are produced by systematically releasing water from large tanks.

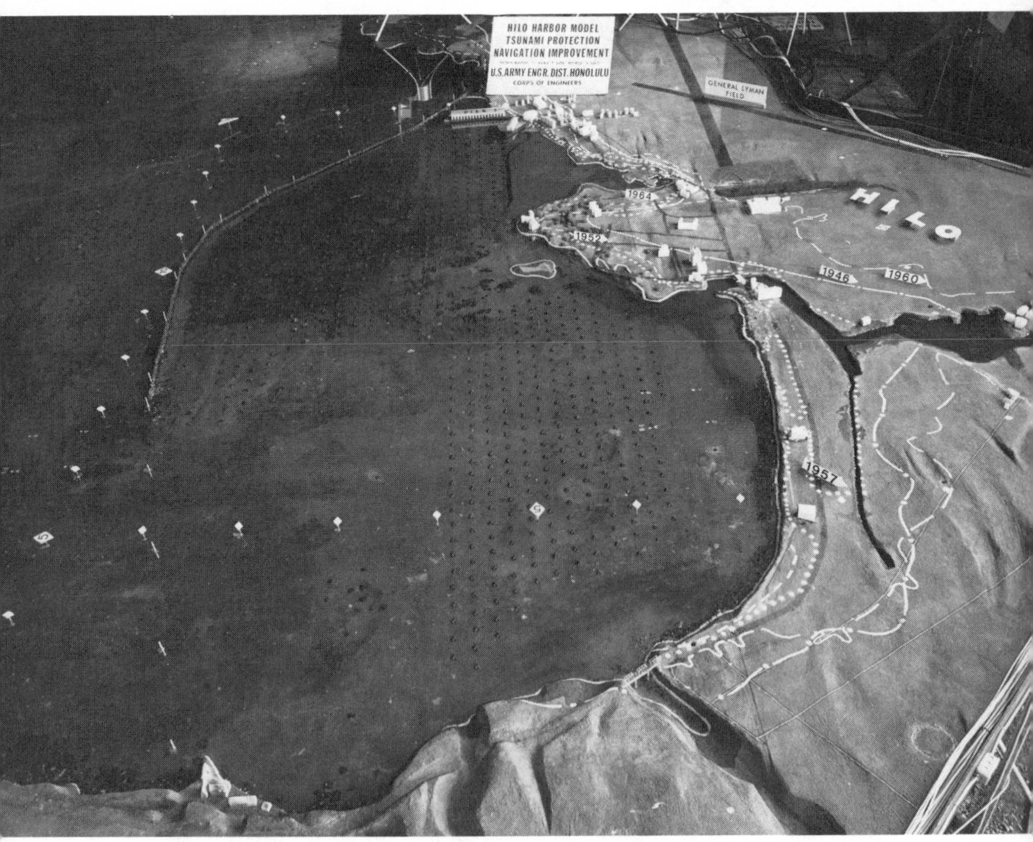

Figure 8.3 Hydraulic model of Hilo Bay. The horizontal scale is 1:600 and the vertical scale is 1:200. The lines shown on the model indicate the limits of inundation of the tsunamis of 1960 (dashed line), 1957 (dotted line), and 1946 (dashed and dotted line). The commercial docks can be seen at the top of the picture. The section of the breakwater just to the left of the docks was carried away during the 1946 tsunami. The railroad bridge across the Wailuku River can be seen in the lower right-hand corner of the picture. The bridge was destroyed during the 1946 tsunami.

Experiments with this hydraulic model have shown that almost any wave that enters the bay directly hits downtown Hilo or is bounced off the Hamakua coast into the town. Sometimes the direct and reflected waves interact to produce especially large waves in the center of the bay near Cocoanut Island.

In today's computer age, another type of model is being used increasingly to study tsunamis. This technique, called a "numerical model," uses a high-speed digital computer to calculate mathematical simulations of tsunami waves (Figure 8.4).

These models, however, are only as good as the data upon which they are based. These data can come only from measurements and observations of real tsunamis.

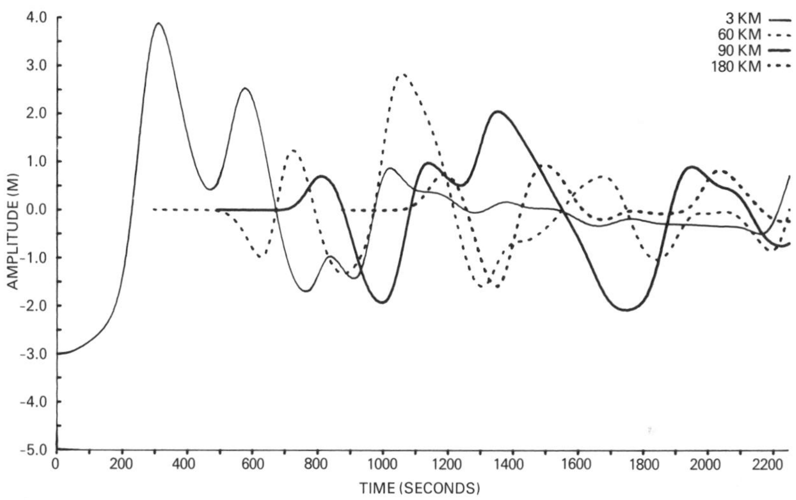

Figure 8.4 Graphic representation of shoreline wave heights from a mathematical model of the 1975 Hawaiian tsunami.

Advancements in the Warning System

Since its inception in 1948, the Warning System has sought constantly to improve the reliability and accuracy of its operations. A recent advancement has been the use of satellites to transmit data from tide stations.

Satellite-telemetered tide stations have been installed at some 25 sites across the Pacific. These stations operate on their own independent power sources, secure from electricity outages resulting from earthquakes. Sea-level measurements are made every two seconds, averaged over a three or four minute interval, and the data are transmitted by satellite to the PTWC every three to four hours. In the event of a tsunami wave, however, an "event detector" almost instantaneously sends a message to the PTWC over a special emergency satellite channel. Although satellite coverage is limited at present to the southwest and south Pacific (including South America), future satellite data-collection platforms will be used to help cover the entire Pacific region. The new satellite-telemetered tide stations go a long way in helping to fill the gap in confirming the generation of tsunamis from the coast of South America.

But even more sophisticated measuring devices are being tested, devices that do not have to be installed on land. A warning system has been proposed that would be based on the detection of tsunamis by highly sensitive bottom-pressure gauges located on the seafloor in mid-ocean. According to specialists, a tsunami wave as small as a quarter of an inch could be detected with such instruments. And, in fact, a small tsunami resulting from a Mexican earthquake in March of 1979 was successfully measured using a device on the seafloor off Baja, California, nearly 10,000 feet deep. To study how tsunami waves change upon entering shallow water, an observational program has been conducted off the Galapagos Islands, with the instruments installed at depths of 10,000 feet, 33 feet, and 3 feet.

In addition to seafloor tsunami detectors, seismographs are now being deployed on the ocean bottom. The Japanese have been successfully operating a permanent ocean bottom seismograph system off the southern coast of central Honshū. Attached to their seismographs is a tsunami gauge.

At the present time, the Pacific Tsunami Warning Center directly monitors telemetered tide gauges around the Hawaiian Islands and makes use of an extensive, sophisticated communications network to contact the seismic observatories, tide stations, and points for the dissemination of information around the Pacific basin. State-of-the-art, high-speed computers are used to perform the complex calculations necessary to predict the arrival time of tsunami waves. In the future, the system will ultimately use both shore-based and ocean-bottom earthquake and tsunami sensors transmitting real-time data to the center via satellite, where the computers will provide the scientists with the information they need to issue confirmed tsunami alerts.

Recent Tsunamis

Even though the Hawaiian Islands have not been assaulted by a major Pacific-wide tsunami for more than 25 years, tsunamis continue to be a real and present danger in the Pacific region.

Just before 9 A.M. local time on May 26, 1983, a large earthquake, measuring 7.7 on the Richter scale (Figure 8.5), occurred in northern Japan. At 9:14 A.M. the Japanese issued a tsunami warning of the highest degree, "Great Tsunami," for coastal areas on the Sea of Japan in northern Honshū. But the first tsunami waves had already reached the coast of Aomori and Akita prefectures by 9:08 A.M. More than one hundred people were killed as waves reportedly ran up more than 30 feet along the shores of Akita, Aomori, and Hokkaido.

Figure 8.5 Seismogram of the May 6, 1983 Japanese earthquake, measuring 7.7 on the Richter scale, recorded at the Honolulu Observatory of the Tsunami Warning System.

Waves continued to crash ashore throughout the night and well into the next day as the tsunami was repeatedly reflected back and forth across the Sea of Japan. Even two days later small tsunami waves could be recognized at some tide stations.

The fishing industry suffered heavy loss with more than 250 boats sunk outright and some 1,600 vessels washed away or damaged. Crop fields were flooded and many houses destroyed, resulting in damage totalling $800 million.

On March 3, 1985 a major earthquake again struck Chile. Measuring 7.4 on the Richter scale, the epicenter lay off the coast near Valparaiso. The earthquake claimed 124 lives and more than 2,000 people were injured in Valparaiso and Santiago, Chile. A small tsunami was generated, which measured less than 4 feet at Valparaiso but advanced across the Pacific and was recorded at many tide stations of the Tsunami Warning System. At Papeete, Tahiti, the tsunami waves measured 4 inches, and just over 1 inch at Honolulu. But Hilo, always sensitive to tsunamis, recorded waves nearly 20 inches high, second only to Valparaiso.

The great Mexican earthquake of 1985 also generated a small tsunami. At 6:18 A.M. on September 19 the main earthquake, measuring 8.1 on the Richter scale, struck Mexico. Centered about 250 miles southwest of Mexico City, the quake had a devastating effect on the heavily populated metropolis, resulting in a death toll mounting to 10,000 and leaving over 600,000 homeless. The tsunami generated by this earthquake spread out from its source across the Pacific. At the coastal town closest to the epicenter, Lazaro Gardenas, the largest waves were about 9 feet high and ran ashore as far as 180 feet inland. Along the Mexican coast from Manzanillo to Acapulco the tsunami waves ranged from 3 feet to about 10 feet high. At Hilo the waves measured 9 inches and at Tahiti the tsunami was only 2 inches high. How is it that such a giant earthquake generated such a small tsunami? Scientists believe that the continuing collision of the giant plates that make up the earth's surface around Mexico produces little vertical movement of the seafloor, and hence little displacement of the sea water above the site of the earthquake.

Another Aleutian Tsunami

On May 7, 1986 at 2247 GMT, an earthquake measuring 7.6 on the Richter scale struck the Aleutian Islands. The epicenter was located about 100 miles southeast of Adak Island and approximately 70 miles southwest of Atka Village, Alaska. Less than 30 minutes later, the Pacific Tsunami Warning Center had sent out messages to tide stations requesting confirmation of a tsunami. The Adak tide gauge station, where a wave nearly 6 feet high was registered, confirmed the generation of a tsunami. A Regional Tsunami Warning was issued by the Alaska Tsunami Warning Center at 2315 GMT, and a Pacific-wide Tsunami Warning was issued at 2351 GMT (1:51 P.M. Hawaii time) by the Pacific Tsunami Warning Center.

The epicenter of the earthquake was very close to that of the 8.3 magnitude earthquake on March 9, 1957, which had generated a Pacific-wide destructive tsunami (described in Chapter 4). A report came in from the Midway Island tide station which recorded waves over 2 feet high—similar to those wave heights registered at Midway during the 1957 tsunami. Concern grew as the waves headed for the main Hawaiian Islands.

The earthquake had, indeed, generated a Pacific-wide tsunami, which would be recorded at tide stations throughout the Pacific, but a large destructive tsunami failed to materialize. At 0510 GMT the Pacific Tsunami Warning Center cancelled the warning, with predictions that maximum sea level fluctuations of 50 centimeters (about 20 inches) would be measured at some tide stations. Although there was no damage, Honolulu recorded 40 centimeters; Hilo, 55 centimeters; Hanalei, Kauai, 61 to 91 centimeters; and Kapaa, Kauai, up to 122 centimeters.

Both the Alaska Tsunami Warning Center and the Pacific Tsunami Warning Center had done an excellent job in providing a timely tsunami warning to the Pacific basin. The Tsunami Warning was issued only 64 minutes after the earthquake itself.

Although the warning system operated smoothly, the Civil Defense alert and evacuation procedures did not go so well. The Emergency Broadcast System failed to work; incredible as it seems, there had been no regular testing procedures. Evacuation efforts, however, did result in thousands of beachgoers, tourists, and workers in coastal areas moving inland. Hundreds of yachtsmen took their boats to sea where they could be seen off Waikiki. Police tried to keep the evacuation moving smoothly, but by 4 P.M. most of the main streets in Honolulu had become hopelessly jammed as the normal afternoon rush hour got an early start with the closing of state and county offices and private businesses. In some areas, residents simply refused to leave the seashore—continuing to swim, surf, and fish.

Even in quiet Hilo traffic was snarled. And just as in 1960, there were Hilo residents who refused to believe there was anything to fear; it was "tsunami party time!" On Oahu it was reported that a solid line of traffic was actually headed toward Waikiki just before the first wave was due to strike!

Once again, radio disc jockeys added to the confusion with unofficial and inaccurate reports of wave activity. The international press reported that there had been "No tsunami!" Indeed, many people viewed the alert as another false alarm; fortunately, no great waves came crashing ashore in the Hawaiian Islands or there would certainly have been many casualties. But the Warning System had responded to the large earthquake in the only way possible, with prudence and caution. Far better that people be evacuated from threatened areas and have only small tsunami waves arrive, than be caught unaware by giant waves like those of 1946 that brought horrible deaths and enormous destruction.

Perhaps the May 7, 1986, tsunami should best be viewed as an exercise. Government agencies desperately need to learn from the experience and be better prepared for the next tsunami.

The Next Tsunami

Some people wonder about the odds of another big tsunami in Hawaii. If we look at the historical records kept since 1813, we find a total of 95 tsunamis in 175 years, or about one tsunami every two years. There has been no really large tsunami since 1960. Does this mean that we are long overdue? Well, events like earthquakes and tsunamis cannot be predicted by such elementary statistics. What we can be sure of is that another giant tsunami will come one day. It may not strike tomorrow, or next week, or even next year, but there will be another tsunami and more great waves will crash ashore in Hawaii. We will probably have adequate warning. Will the warning be understood and heeded? That is up to you.

REFERENCES

Adams, W. M. and N. Nakashizuka. 1985. A working vocabulary for tsunami study. *Tsunami Hazards.* 3:45–51.

Bennett, C. C. 1869. *Honolulu Directory and Sketch of the Hawaiian or Sandwich Islands.* Honolulu: Honolulu Press.

Bingham, H. 1847. *A Residence of twenty-one Years in the Sandwich Islands.* 3rd ed. Canandaigua: Goodwin.

Bonk, W. J., R. Lachman, and M. Tatsuoka. 1960. A report of human behavior during the tsunami of May 23, 1960. Unpublished report of the Hawaiian Academy of Sciences, Hilo.

Coan, T. 1882. *Life in Hawaii.* New York: Randolf.

Cox, D. C. and J. Morgan. 1977. Local tsunamis and possible local tsunamis in Hawaii. Hawaii Institute of Geophysics report no. 77–14.

Dana, J. D. 1868. A letter of the School Inspector General in *Hawaiian Gazette,* April 29, 1868. *American Journal of Science.* Series II, 46:105–123.

Eaton, J. P., D. H. Richter, and W. U. Ault. 1961. The tsunami of May 23, 1960, on the island of Hawaii. *Bulletin of the Seismological Society of America.* 51:135–157.

Fraser, G. D., J. P. Eaton, and C. K. Wentworth. 1959. The tsunami of March 9, 1957, on the island of Hawaii. *Bulletin of the Seismological Society of America.* 49:79–90.

Hitchcock, C. H. 1911. *Hawaii and Its Volcanoes.* 2nd ed. Honolulu: The Hawaiian Gazette Co., Ltd.

Jaggar, T. A. 1946. The great tidal wave of 1946. *Natural History.* 55:263–268, 293.

Macdonald, G. A., F. P. Shepard, and D. C. Cox. 1947. The tsunami of April 1, 1946 in the Hawaiian Islands. *Pacific Science.* 1:21–37.

Malo, D. 1951. *Hawaiian Antiquities.* Honolulu: Bishop Museum Press.

Moore, J. G. and G. W. Moore. 1984. Deposit from a giant wave on the island of Lanai, Hawaii. *Science.* 226:1312–1315.

Whitney, H. M. 1868. On the earthquake and eruptions of 1868. In J. Dana, ed., Recent eruptions of Mauna Loa and Kilauea. *American Journal of Science.* Series II, 46:112–115.

ILLUSTRATION CREDITS

Anonymous: Figure 1.10

James W. Duncan: Figure 1.3

Joe Halbig, University of Hawaii—Hilo: Figure 7.4

Hawaii Tribune-Herald: Figure 5.2

Henry Helbush: Figure 4.1

Joint Institute for Marine and Atmospheric Research, University of Hawaii—Manoa, courtesy of George Curtis: Figures 1.12, 1.13, 4.2, 8.2

Joint Tsunami Research Effort: Figure 1.6

Look Laboratory: Figure 8.3

Ted Lusby: Figure 1.5

Gordon A. Macdonald, Agatin T. Abbott, and Frank L. Peterson, *Volcanoes in the Sea: The Geology of Hawaii* (second edition), pp. 318–319, copyright 1983 by University of Hawaii Press: Figures 2.6, 6.1

Charles L. Mader: Figure 8.4

Magoon Private Collection: Figures 1.8, 1.11

National Oceanic and Atmospheric Administration (NOAA)/EDIS: Figures 1.2, 5.1, 6.2, 6.3, 6.4, 7.5

Pacific Tsunami Warning Center (U.S. Department of Commerce/NOAA): Figures 3.1, 3.2, 3.4, 8.1, 8.5

Pacific Tide Party: Figure 5.5

University of California—Berkeley: Figures 1.4, 1.9, 2.5, 2.7

U.S. Army Corps of Engineers: Figures 1.7, 5.4

U.S. National Park Service: Figures 7.2, 7.3

INDEX

If you are one of those who have experienced the terror of tsunami waves and would like to have your story recorded, we would be interested in hearing from you. Please write to the following address:

Dr. Walter C. Dudley
Natural Science Division
University of Hawaii at Hilo
Hilo, Hawaii 96720-4091